ROUTLEDGE LIBRARY EDITIONS:
MONETARY ECONOMICS

Volume 5

KEMMERER ON MONEY

KEMMERER ON MONEY

EDWIN WALTER KEMMERER

Routledge
Taylor & Francis Group
LONDON AND NEW YORK

First published in 1934 by George Routledge & Sons, Ltd.

This edition first published in 2017
by Routledge
2 Park Square, Milton Park, Abingdon, Oxon OX14 4RN

and by Routledge
605 Third Avenue, New York, NY 10017

Routledge is an imprint of the Taylor & Francis Group, an informa business

British Library Cataloguing in Publication Data
A catalogue record for this book is available from the British Library

ISBN: 978-1-138-73264-3 (Set)
ISBN: 978-1-315-16457-1 (Set) (ebk)
ISBN: 978-1-138-63453-4 (Volume 5) (hbk)
ISBN: 978-1-315-20673-8 (Volume 5) (ebk)

Publisher's Note
The publisher has gone to great lengths to ensure the quality of this reprint but points out that some imperfections in the original copies may be apparent.

Disclaimer
The publisher has made every effort to trace copyright holders and would welcome correspondence from those they have been unable to trace.

KEMMERER ON MONEY

BY

EDWIN WALTER KEMMERER

LONDON

GEORGE ROUTLEDGE AND SONS, LTD.

BROADWAY HOUSE: 68–74 CARTER LANE, E.C.

1934

TO THE MEMORY OF
GROVER CLEVELAND
A FEARLESS AND VALIANT
CHAMPION OF SOUND
MONEY

PREFACE

PROBABLY no economic institution of our modern world is affected with a greater public interest than money. In terms of money, practically all of the world's production and trade is carried on, all its wages paid, and all its debts are expressed, carried, and liquidated. The modern man is money-minded and most of the values he contemplates in his daily life are money values.

In times of economic disorder, therefore, the man in the street jumps to the superficial conclusion that the cause of the trouble is a breakdown in this all-pervasive thing, money, and that the remedy is to be found in some kind of monetary reform. Furthermore, since the average man measures his wealth and, to a large extent, his welfare in terms of the amount of money he commands, and since no man has "enough money" to satisfy himself, and nearly everyone finds his supply of money reduced in times of economic depression, the conclusion is natural that, if the country only had more money, all would be well. It is significant in this connection that, just before Germany in the autumn of 1923 stabilized her inflated paper mark at the astronomical figure of one trillion paper marks to one gold mark, there was widespread complaint in Germany that there was not enough money in circulation.

Popular familiarity with money in its superficial aspects makes everyone "a monetary expert". There is probably no other economic institution on which so many people speak "with vigour and authority" as money, and no other field of economics can boast of a mass of barnacles so bright in appearance and with such clinging propensities as "the money cranks" in the field of monetary science—men who solve all our economic problems by monetary panaceas which they have concocted without being handicapped by any knowledge of the elementary principles of economics or of the facts of monetary history. Few people who have never studied the principles of chemistry or biology would venture to propose schemes for the offhand solution of the most complicated problems in those subjects, but, when it comes to the equally highly complicated subject of money, we have "authorities" by the million.

Because so much depends upon the value of the monetary unit and the orderly functioning of the monetary system in a modern state, it is exceedingly important that public confidence in a country's money should be maintained, and this confidence is exceedingly vulnerable to attacks from those in high political positions, either administrative or legislative.

Monetary science is a branch of economics, and the student of money is largely concerned with the application of economic principles to the special subject of money. A knowledge of fundamental economic principles is essential to even an elementary understanding of money, and there is needed in addition, of course, a knowledge of the principles of money itself.

For generations many of the best minds of the world have been devoted to the study of the science and art of money. All sorts of theories, sound and unsound, have been advanced, subjected to merciless criticism by men specializing in monetary science throughout the world, and have been thrown into the monetary scrap-heap, or have been adjusted and adapted and passed on as accepted, or at least tentatively accepted, principles in a developing science of money. These principles have been continually submitted to the test of monetary experience in a world in which nearly every country for centuries has been experimenting with money. As a result of these years of study of monetary principles and of monetary history, there has evolved a large body of monetary knowledge which we call monetary science, and, while there is still an immense amount of work to be done in this dynamic field, and while there is much difference of opinion among monetary scientists on matters of detail, as there is among scientists in every other field, a widespread agreement exists to-day among monetary scientists concerning the fundamental principles of money and the outstanding lessons of monetary history.

There are in the world at the present time hundreds of men, of whom the author of this book is one, who have devoted their lives to the study of money and monetary problems, with the sole desire of arriving at the truth and of making that truth available to a vitally interested public. In times like these, it is the political duty of these men to express publicly and in no uncertain language their opinions on monetary problems to a world that is desperately groping for the truth as a basis of monetary policy.

This is the excuse for this little book, which discusses twelve monetary topics of current importance in the United States to-day. The chapters constitute a revision and amplification of twelve articles which the author contributed to the *New York Sun.*

EDWIN WALTER KEMMERER

PRINCETON UNIVERSITY,
PRINCETON, NEW JERSEY,
U.S.A.

CONTENTS

CHAPTER I

THE GOLD STANDARD

What It Is—Its Principal Defect—The Only Standard that offers an Early Hope of being an International Standard——The Problem of To-day is not to create a Substitute for the Gold Standard, but to make the Gold Standard a Better Standard

EVERYBODY to-day is interested in the gold standard. Apart, however, from a small group of economists, few people could answer the question : " Exactly what is meant by the gold standard ? " Even the economists themselves would differ concerning the precise definition—as do thc specialists in any subject concerning the exact definition of general terms—but they would agree substantially in fundamentals. What is this gold standard of which we are now hearing so much, both of praise and blame ?

What is the Gold Standard ?

Briefly, the gold standard is a monetary system in which the unit of value, be it the dollar, the pound, the franc, or some other unit in which prices and wages are customarily expressed and in which debts are usually contracted, consists of the value of a fixed quantity of gold in a free gold market. In this definition several things that are popularly associated with the gold standard are conspicuous by their absence. There is no mention of gold coins, of legal tender, of free coinage, or even of redemption of paper money in gold. These things are all customary accompaniments of the gold standard. They are useful devices for maintaining it. The gold standard, however, could exist without any or all of them. Furthermore, currency might have all these attributes and still not be a true gold standard.

In 1919 and early 1920, for example, in the Union of South Africa, gold sovereigns, which were unlimited legal tender and which enjoyed the free coinage privilege in England, circulated freely and the country's bank notes were redeemable at the banks of issue in gold on demand at legal parity. But, since the exportation of gold was rigidly restricted by the Government, there was no free market for gold, and since South African gold coins could not be exported to what would otherwise have been their best market, they were dammed up in the country.

A sovereign in South Africa, therefore, was worth much less than it was worth in the free gold markets of the world. In order to get the sovereigns with which to redeem their notes on demand, as required by law, the South African banks of issue were compelled to buy raw gold in London at a premium, to get it coined in London in the usual way, and then bring it to South Africa. At times they had to pay as much as 26 or 28 shillings of South African paper money to obtain a sovereign in England. The sovereign was then paid out in South Africa by the bank of issue in redemption at par of a 20-shilling bank note.[1] When the gold value of money within a country is thus artificially held down by Government action, the country cannot be said to be on a true gold standard even though its legal tender paper money is convertible into gold coin at par on demand.

Varieties of the Gold Standard

There are several varieties of the gold standard. There is, for example, the gold coin standard, which was the monetary standard of the United States prior to the spring of 1933. Here all the different kinds of money were kept at a parity with gold coins in a gold market that was absolutely free.

[1] See Kemmerer-Vissering, *Report on The Resumption of Gold Payments by The Union of South Africa*, 1925, pp. 536–7.

Another variety is the gold bullion standard, the monetary standard that England had from the time of her return to the gold basis in 1925 until the breakdown of the gold standard in September, 1931. Under this standard there was practically no gold coin in circulation, but all the money of the country was maintained at gold parity. To this end the Bank of England redeemed its notes on demand in gold bars whenever gold was needed for making international payments.

Then, again, there is the widely used gold exchange standard, under which frequently there is no gold coin in circulation and no redemption of paper money in either gold coin or gold bars. Here the various kinds of money within the country are redeemable in drafts drawn on gold deposits held abroad.

People often contrast the gold standard currency system with a managed currency system and, in fact, the two types of currency are usually fundamentally different. However, all gold standard currencies are more or less "managed" through the discount and open-market operations of central banks and other central bank devices. Furthermore, although it would be a dangerous experiment to try, and one that for political and social reasons would probably soon break down,[1] it would be theoretically possible, from a purely economic point of view, so to regulate the supply of paper money and bank credit within a country as to maintain the value of the money unit at home at approximately the value of a fixed quantity of gold in the free markets of the world, without any system whatever of demand interconvertibility of paper money with gold. In other words, a highly managed paper money currency might conceivably be on the gold standard.

Great Britain's inconvertible managed paper pound was maintained so nearly at a fixed gold value during the first six months of 1933 that it was not far from being at that time a managed paper currency on the gold standard.

America's Gold Coin Standard

Since the principles underlying the different varieties of the gold standard are the same, and since the gold coin standard of the United States of the type existing prior to March, 1933, has been the most widely used type of the gold standard in modern times, the nature of the gold standard may well be illustrated by reference to the American experience.

In the United States the unit of value was

[1] See author's "Controlled Inflation", in *American Economic Review, Supplement,* March, 1934.

the gold dollar, which is perhaps best thought of as the equivalent of one-tenth of a ten-dollar gold piece. The weight of the ten-dollar gold piece was 258 grains, of which 232·2 grains were gold and the balance was alloy, consisting chiefly of copper. Since there are 480 grains in an ounce and since the gold dollar contained 25·8 grains of American "standard" gold, namely, gold nine-tenths fine, and since, therefore, an ounce of standard gold would coin into $\frac{480}{25 \cdot 8}$, or $18·60, anyone could take American standard gold in any quantity to an American mint and have it minted into gold at this rate. In other words, anyone could get $18·60 of gold coin (less certain petty charges for assaying and refining) for each ounce of standard gold he took to the mint; and, conversely, anyone could melt down full-weight standard gold coins and obtain from them approximately an ounce of American standard gold for each $18·60 of coin melted. Thus an ounce of standard gold in the United States was always equal in value to $18·60. No matter how much gold or how little gold was being produced in the world, this price was fixed and invariable. In fact, $18·60 could be thought of as merely the name of an ounce of American standard gold in the form of money. To say that this price never changed was a truism; it was like saying: "A foot is always twelve inches long."

Value of all Money tied to Gold

At the end of February, 1933, just before the United States went off the gold standard, there were in America a little over eight billion dollars of money (aside from that held in the United States Treasury). Of this sum about two and a third billion dollars consisted of gold coins and gold certificates. The rest was made up of six different kinds of paper money, the gold value of all of which was guaranteed by the United States Government, and of various coins of silver, nickel, and copper which the Government was required by the Gold Standard Act of 1900 to maintain at a parity with gold coin. All the dozen different kinds of money, of which there were in circulation at that time about six and a half billion dollars, were in practice interchangeable on demand at par with gold throughout the country. The value of a dollar of each kind of money was, therefore, tied to the value of a dollar of gold.

Deposit Currency tied to Gold

When the United States was on the gold standard, not only did the value of all its money rise and fall with changes in the value of gold, but the same was true of the values of all its bank deposits which were payable in this money and which even at the end of the depression year 1932 amounted to about forty-two billion dollars. The money in hand-to-hand circulation in the United States probably circulates on an average somewhere between thirty to forty times a year, while every dollar of gold held in the Federal Reserve banks, as reserve money for bankers' deposits under the gold standard, did something like forty times as much money work as it would have done in hand-to-hand circulation. This work was done through the circulation of bank cheques, by which about 90 per cent. of the total business is done in the United States—a business which in a prosperous year probably amounts to over a trillion dollars. For the depression year 1932 the demand deposits of American banks in 141 leading cities had an average velocity of circulation [1] of about 33. In other words, for each dollar of demand deposits held on the average throughout the year there was performed $33 of money work by means of cheques. The value, moreover, of every dollar of these hundreds of billions of dollars of business transactions was

[1] A discussion of the concept of "Velocity of Circulation" is given on pp. 149–53.

equivalent to the value of 25·8 grains of American standard gold: namely, to the gold standard dollar.

Debts and the Gold Standard

The total volume of internal interest-bearing debts outstanding in the United States in 1932, which Evans Clark in his book, *The Internal Debts of the United States*, estimates at about two hundred and eighty-five billion dollars, was payable, while the United States was on the gold standard, in dollars whose value was maintained always at the equivalent of 25·8 grains of American standard gold. When gold went up in value for any reason, the value of the United States dollar likewise went up, commodity prices fell, and the debtor had to pay the creditor in settlement of his debts dollars of increasing purchasing power, although he usually received fewer dollars than formerly for the things he had to sell. Here the debtor suffered.

When, on the other hand, gold went down in value, the value of the dollar likewise declined, commodity prices rose, and debt burdens were all made lighter because the debtor received continually increasing prices for the things he had to sell and he paid the creditor in cheaper dollars the amount of his original obligation. Here the creditor suffered.

Obviously, in the interest of justice to both debtor and creditor, it was desirable that the value or purchasing power of this gold monetary unit should be stable. Due, however, to the great variations in the world's supply of gold from time to time, and equally to the frequent great changes in the world's demand for gold, the value of gold when viewed over any considerable periods of time has been unstable. If we think of the United States gold dollar during the years 1896 to 1932 as a yardstick of value and represent the purchasing power of this dollar over commodities and services in general in the year 1913 by a length of 36 inches—namely, a yard—the length of this yardstick of value would have been as follows for the dates specified:

1896	51 inches
1913	36 ,,
1920	19 ,,
1922	23 ,,
1926	21 ,,
1929	20 ,,
1932	27 ,,
1933	28 ,,

These wide variations in the yardstick of value, in terms of which practically all debts were contracted and paid, were the most serious defect of the gold standard. The same defect, however, was also found in the silver standard, the bimetallic standard, and every other metallic standard with which the world has ever had experience. While during some periods in history silver has been more stable in value than gold, since 1890 it has been somewhat less stable. Furthermore, the world's experiences to date with so-called " managed paper money " standards, when they have been tried for any appreciable number of years, have shown them to be even more unstable in value than gold.[1]

The Dollar Stable in Value, 1921–29

It is a significant fact, moreover, upon which proponents of the gold standard place much emphasis, that, after the early post-war deflation, the gold dollar showed a remarkable stability in value as measured by general prices in the United States for about nine years. In fact, from 1921 to 1930, inclusive, the extreme average annual variation in this yardstick of value was only about 3 inches: namely, from approximately 23 inches in 1922 to 20 inches in 1929. During the economic depression, however, gold (like every other important basic

[1] The subject of inflation and deflation in relation to debts, with particular reference to paper money standards, is discussed later. See Chapter IX.

commodity) has shown wide variations in value, but there is strong evidence that the causes of these variations and of the present high value of gold are of a temporary character.[1]

Improvement of Gold Standard rather than Displacement Desirable

Before the Great War most of the world was on the gold standard. During the years 1914 to 1919 this standard was practically everywhere suspended. The paper money systems of these years, however, proved universally unsatisfactory, and shortly after the war the tide again turned toward the gold standard. By 1929 most of the advanced countries of the world were again on the gold basis. Then came the crash of October, 1929. Since then the world depression has forced most countries again on to a paper money basis; but in every case, except that of the United States, when the gold standard was given up, it was given up reluctantly by the Government and under heavy pressure. Nearly everywhere the sentiment to-day is strong for a return to gold as soon as practicable.

It is highly important from the standpoint

[1] See Kemmerer, E. W., " The Gold Standard and the Present Economic Situation," *Proceedings of the American Philosophical Society*, Vol. 71, No. 3, pp. 87–93, 97–102; *infra* pp. 180–3.

of international trade and finance that the principal countries of the world shall all have the same monetary standard, and the gold standard is the only standard which offers an early hope of becoming an international standard.

Although far from being a perfect monetary standard, gold is the best standard with which the world has had any extended experience. The urgent monetary problem of the immediate future is not a *national* problem of creating a substitute for the gold standard, but rather an *international* problem of making the gold standard a better standard. In the words of the Macmillan Report of 1931 :

> There is, perhaps, no more important object within the field of human technique than that the world as a whole should achieve a sound and scientific monetary system. But there can be little or no hope of progress at an early date for the monetary system of the world, except as the result of a process of evolution starting from the historic gold standard.

Chapter II

THE PAPER MONEY STANDARD OF THE UNITED STATES IN 1933

Fundamental Differences between it and the Gold Standard —The Eight Kinds of Money of the United States, their History and Uses, and how they are affected by the Present Federal Policies

Fundamental Differences between a Gold Standard and a Paper Money Standard

In the preceding chapter the nature of the gold standard was described. In the spring of 1933 the United States went on a paper money standard. In what important respects do these standards differ ? The answer is twofold :

First, the value of a paper money dollar, unlike that of the gold dollar, is not tied to a commodity of a high value in the world market. The Government itself, or the central bank which issues the paper money, determines how much shall be issued. Since there is no legal obligation to redeem the paper money in gold or in any other manner to maintain its value at a parity with some generally marketable commodity, " the sky is the limit " of possible issue and, therefore, of possible fluctuations in value.

Second, a paper money standard is a standard that operates only within a very limited area, usually an area not wider than a single nation, although there have been occasions, such as that of the sterling paper money standard of Great Britain at the present time, when essentially the same standard was operating in several countries closely related economically or politically. Ordinarily, however, one sovereign State does not wish to have the value of its paper money unit determined or controlled by another State. With a paper money standard, therefore, each country is usually a law unto itself in determining how much or how little paper money it will issue from time to time in the light of what it considers to be the national economic and political necessity. The usual result is, as it was during the paper money régimes of the World War and early post-war period and as it is to-day, that each country has a different standard, the monetary unit of which rises or falls in value with little or no regard to the fluctuations in the monetary units of other countries. This is, of course, a great handicap to international trade and international finance, for it involves widely fluctuating exchange rates among all paper money standard countries and between each such country and the gold standard countries.

United States Money before and immediately after the Discontinuance of the Gold Standard in 1933

Let us now examine the significant changes that took place in the different elements in the monetary system of the United States between February, 1933, and the passage of the Gold Reserve Act of January 30, 1934. Among the advanced countries of the world the United States was unique in the large number of different kinds of money it had in circulation and in the complexity of its monetary system. In addition to its minor coins of nickel and copper, its fractional silver coins, and to the Treasury notes of 1890, of which only a few remained in circulation, it had eight distinct kinds of money. They were: (1) gold coins; (2) gold certificates; (3) silver dollars; (4) silver certificates; (5) United States notes or greenbacks; (6) national bank notes; (7) Federal Reserve notes, and (8) Federal Reserve Bank notes. Let us briefly consider each of these eight kinds of money, noting, in passing, that all kinds of United States money, including even the fractional silver, nickel, and copper coins, were surprisingly made unlimited legal tender for all debts, public and private, by Joint Resolution of Congress, approved June 5, 1933.

Gold Coins

American gold coins, first authorized by the Mint Act of 1792, and slightly changed in their pure gold content on two occasions about one hundred years ago to restore and maintain their circulation under the bimetallic system of that time, had circulated continually throughout the United States for nearly a century, except during the greenback period from 1862 to 1879 when their circulation was narrowly limited to the Pacific coast. Never in the history of the United States had there been any legal restriction on the holding or payment of gold coins.

With the legislation of March, 1933, all was changed. Gold coins were no longer minted and no longer permitted to circulate in the country, and the mere holding of them was made illegal under heavy penalties. At the current dollar-franc exchange rate a dollar in gold coin was worth about $1·57 of United States paper money at the end of the year 1933. The National Government, however, through its fiscal agencies, was paying foreigners abroad and American miners of new gold at home at that time, approximately $16·47 of paper money for the amount of gold in a ten-dollar gold piece,[1] but was insisting under heavy penalties that everyone in the United States owning American gold coin, so-called " gold hoarders ", should sell it to the Government at the rate of one paper dollar for each dollar of gold coin.

Gold Certificates

The gold certificates, or " yellow backs " as they were popularly known, date back to 1863, when a law was passed authorizing the Secretary of the Treasury to issue them against deposits at par of gold coin and bullion, the gold " deposited for or representing the certificates " to be " retained in the Treasury for the payment of the same on demand ", and " to be used for no other purpose ". They were always looked upon by the public as a form of money partaking of the nature of warehouse receipts. The gold belonged to the owner of the certificate and " it circulated by proxy " in the form of the certificate.

These gold certificates, however, are no longer issued and to hold them to-day is illegal. Moreover, even if a person holding them could demand and receive for them the gold they represent, immediately upon its receipt the gold

[1] A discussion of the Government's gold purchase policy will be found in the next chapter.

itself could be taken away from him and he would be subject to a penalty equal to twice the value of the gold he received. A new form of "gold certificate" was authorized by the Act of January 30, 1934. It is not convertible on demand into gold and may be held only by the Government and the Federal Reserve banks.[1]

The Standard Silver Dollar [2]

The silver dollar was first authorized under the Mint Act of 1792, and it has to-day the same pure silver content that was given it by that law. Because of its inconvenient size, the silver dollar has for many years been an unpopular coin and the number of silver dollars in circulation has long been less than 1 per cent. of the total monetary circulation. Being maintained by the Government at par with the paper dollar, the silver dollar was worth, approximately, 64 cents gold near the end of the year 1933, and since the value of the silver in the dollar in the free markets of the world was then only about 34 cents paper money, the silver content of the dollar was then a negligible factor in maintaining its value, and it continues to be so.

[1] See pp. 193–4.

[2] A more detailed discussion of silver and the silver dollar will be found in Chapters VII and VIII.

Under the Agricultural Adjustment Act of May, 1933, the President was authorized by mere proclamation to determine upon a mint ratio between gold and silver coins and to " provide for the unlimited coinage of such gold and silver at the ratio so fixed ", under certain limitations so broad as to have little restrictive force. Despite the fact that there has been no bimetallism in the world since 1873, the President now holds full authority from Congress to open the mints to the free and unlimited coinage of silver dollars and to launch the country upon either a national or international bimetallic experiment. Furthermore, although the market ratio of gold to silver in late December, 1933, was approximately 74 to 1, many of the " silver group " in Congress and of the partisans of silver outside of Congress appeared then, as now, to favour the old bimetallic ratio of 16 to 1.

The President's Proclamation of December 21, 1933, authorized large additional issues of silver dollars.[1]

Silver Certificates

Silver certificates are in all essentials like the gold certificates previously described. They were originally authorized in 1878 and have

[1] See pp. 118–20.

long constituted less than 10 per cent. of the country's total monetary circulation. Inasmuch as the silver back of them is worth in the free markets of the world only a minor fraction of their gold value, its influence in maintaining their value is negligible. Rarely are silver certificates presented for redemption in silver dollars. Silver certificates continue to circulate freely and are still redeemable in silver dollars. Further increases in their issue were authorized under the provisions of the Agricultural Adjustment Act of May 12, 1933, which permits the President to allow countries indebted to the United States under interallied debt agreements to pay up to $200,000,000 in silver at prices well above the market. Against the silver so received the Secretary of the Treasury is required to issue silver certificates. Additional authority to issuc silver ccrtificates was given by the Gold and Silver Reserve Acts of 1934.

The Greenbacks

The fifth important kind of money circulating in the United States is the " greenback ", or United States note. It was the money that constituted the depreciated paper money standard from 1862 to 1879. Greenbacks are issued directly by the Government and are the money that is primarily secured by the $150,000,000 gold reserve established in the Treasury by the Gold Standard Act of 1900. Their circulation has been limited to approximately $346,000,000 since 1878. Being Government paper money with an unfortunate history of depreciation and instability covering seventeen years, from 1862 to 1878,[1] and being a type of money that is particularly likely to be expanded for fiscal purposes by a Government that is hard pressed for funds, these greenbacks have long been considered by students of money to be a weak element in the United States monetary system, and for this reason there has been much agitation for their complete retirement.

Legislation in the spring of 1933 discontinued their convertibility into gold on demand and actually authorized the President to increase the volume outstanding up to three billion dollars, a sum equal to about seven times the maximum amount outstanding at any time in their entire history.

National Bank Notes

National bank notes date back to the National Bank Act of 1863 and are issued by upwards of 5,000 national banks scattered throughout the

[1] See Chapter V.

country. They are obligations of the United States Government, as well as of the issuing banks, and are secured dollar for dollar by a deposit with the Treasurer of the United States of Government bonds and lawful money, and by a 5 per cent. lawful money redemption fund. They cannot be issued in excess of the issuing bank's paid-up capital. They constitute about one-sixth of the country's total monetary circulation, and tend to increase in times like the present when the national debt, by which they are secured, is growing rapidly.

Federal Reserve Notes

By far the most important kind of paper money in circulation in the United States at the present time is the Federal Reserve note, which constitutes over one-half of the total monetary circulation. Federal Reserve notes are obligations of the twelve Federal Reserve banks and of the National Government, and are secured by a specific pledge of gold certificates, high-grade business paper, United States Government securities, and promissory notes collateraled by such securities, altogether amounting to considerably more than the total amount of Federal Reserve notes outstanding. Under the law, as it existed until recently, the notes were all convertible into gold on demand, and the Federal Reserve banks were required to maintain at all times a gold reserve equal to at least 40 per cent. of the notes in circulation, subject to the provisos that, when the gold reserve fell below 40 per cent., the Federal Reserve Board should impose a graduated tax on the reserve deficiency which should increase progressively as the reserve declined, and that the equivalent of this tax should be added to the discount rates charged by the Federal Reserve banks to their customers. This tax and these advances in the discount rate were expected to place a heavy drag on credit expansion at times when declining reserves were danger-signals of inflation.

Under recent legislation, however, these protective features have been weakened and the position of the Federal Reserve notes has been greatly changed. The notes are no longer convertible into gold on demand. The Federal Reserve Board, as heretofore, can suspend from time to time all reserve requirements of the law to any extent it may desire, but now, in addition, it is freed from all obligations to impose a reserve deficiency tax or to require an increase in discount rates as the reserve percentage declines. In other words, there is now no limit whatsoever short of zero beyond which the

Federal Reserve Board cannot, in its discretion, reduce the reserve requirements of the Federal Reserve banks.

Federal Reserve Bank Notes

The last kind of money to consider is the Federal Reserve Bank note. Although these notes were authorized by the Federal Reserve Bank Act of 1913, they have been of little importance until lately, except for a few years, beginning in 1918, in connection with the operation of the Pittman Silver Act. They are in most respects like the bond-secured national bank notes, but they are issued by the Federal Reserve banks themselves. They are obligations of the Federal Reserve banks, and until recently were secured dollar for dollar by United States bonds. No reserve whatever is required against them, except a 5 per cent. redemption fund which must be kept in "lawful money" on deposit with the Treasurer of the United States. At the end of February, 1933, their total circulation was less than three million dollars.

With the legislation of the spring and early summer of 1933, however, these notes loomed into increasing importance. Every month from February to December, 1933, evidenced a rise in their circulation, and by the end of the year the total amount in circulation was $208,000,000. After that date there was a decline.

Under the emergency Banking Act of March, 1933, authority was given to issue these notes not only as formerly, up to 100 per cent. of the United States Government obligations pledged as security, but also up to not more than 90 per cent. of the estimated value of certain specified classes of notes, drafts, bills of exchange, and bankers' acceptances. Since the law places no limit upon the amount of Government debt which the National Government may issue and sell to Federal Reserve banks for these Federal Reserve Bank notes, there is no maximum limit to the possible issue of these notes.

The notes of the German Reichsbank, which after the World War ultimately depreciated to one-trillionth of their original gold value, were very similar to these Federal Reserve Bank notes, and, like them, were secured chiefly by the obligations of the National Government.[1] All things considered, these Federal Reserve Bank notes are probably the weakest link in the country's monetary chain. Being secured chiefly by Government bonds, they provide a possible means of monetary inflation that is peculiarly subtle, because of the widespread

[1] For an account of the depreciation of these German notes, see pp. 79–90.

public confidence in the National Government's promises to pay. Here, although the notes are secured by the bonds, the bonds may be legally paid, principal and interest, in the notes.

The Existence of Radical Monetary Laws an Obstacle to Economic Recovery

In conclusion it should be pointed out that many of the wide powers given to the President in this new legislation are permissive and not mandatory and that a large part of them are of a temporary character granted to enable him to meet emergencies. When this legislation was passed, it was frequently said that the President was given these powers, most of which he would not use, for the purpose of protecting the country from a radical Congress and of protecting Congress from its radical electorates. In a complex situation like this, it is impossible to ascertain and measure motives. One thing, however, is certain: so long as legislation of this kind remains on the nation's statute books and as the President fails to take a strong stand against exercising the radical powers placed in his hands, there will be lacking that confidence which is necessary for a healthy and enduring recovery. In this connection we may well recall the striking words of President Cleveland in his message to a special session of

Congress on August 8, 1893, when the financial foundations of the country were being shaken by the heavy purchases of silver in connection with the monetary experimentation under the notorious Sherman Silver Purchase Act of 1890. He said:

> The people of the United States are entitled to a sound and stable currency and to money recognized as such on every exchange and in every market of the world. Their Government has no right to injure them by financial experiments opposed to the policy and practice of other civilized states, nor is it justified in permitting an exaggerated and unreasonable reliance on our national strength and ability to jeopardize the soundness of the people's money.

CHAPTER III

THE GOLD PURCHASE PLAN

An Attempt to analyse the Plan—Gold is merely a Commodity in a Paper Money Standard Country—England's Experience—Only as a Crude Device for Pumping more Money and Deposit Currency into Circulation does Plan tend to raise Commodity Prices Permanently

The Plan

FOR several months after late October, 1933, the interest of the public both in the United States and abroad in America's monetary experimentation was centred in the Administration's gold purchase plan, a plan that was suddenly and without previous public discussion announced to the world by President Roosevelt in a radio address on October 22.

In this address the President said:

> When we have restored the price level, we shall seek to establish and maintain a dollar which will not change its purchasing and debt-paying power during the succeeding generation. . . .
>
> Our dollar is now altogether too greatly influenced by the accidents of international trade, by the internal policies of other nations

> and by political disturbances in other continents. Therefore, the United States must take firmly in its own hands the control of the gold value of our dollar. This is necessary in order to prevent dollar disturbances from swinging us away from our ultimate goal, namely, the continued recovery of our commodity prices.
>
> As a further effective means to this end, I am going to establish a Government market for gold in the United States. Therefore . . . I am authorizing the Reconstruction Finance Corporation to buy gold newly mined in the United States at prices to be determined from time to time after consultation with the Secretary of the Treasury and the President. Whenever necessary to the end in view, we shall also buy or sell gold in the world market.

The President added that this was " a policy and not an expedient " and that the country was " thus continuing to move toward a managed currency ".

In a word, therefore, the object of the plan was so to control the gold value of the paper dollar as to raise commodity prices to the desired level, presumably that of 1926, and thereafter through currency management to maintain prices at that level.

For the month of October, 1933, in which the radio address was given, the wholesale commodity price level of the United States Bureau of Labour Statistics was about 29 per cent. below that of the year 1926, and the cost of living was about 25 per cent. below.

For reasons in large part, apparently, of a legal character, it was announced a few days later that payment for the gold would be made by the Reconstruction Finance Corporation, not in cash but in its own notes guaranteed by the Government and maturing February 1, 1934, to be "sold to the gold owner at one-quarter of 1 per cent. per annum discount". Of these notes an issue of $50,000,000 was authorized for the purpose of carrying out the plan, and this amount was subsequently increased as needed. In order to assure the "seller" of the gold that his notes were equivalent to cash, the chairman of the Reconstruction Finance Corporation announced that the seller "will have no difficulty in converting his notes into cash at any bank that is a member of the Federal Reserve System for the quoted price for gold on the day that he subscribes for the R. F. C. notes".

The description given above covers in essentials all the public knew of the general nature and purpose of the plan so far as any official explanations were concerned.

Purchases of domestic gold began October 25,

1933, with bids substantially above the current market price, and on October 29 it was announced that the Government purchases would be extended to foreign markets. Foreign purchases were to be made for the account of the Reconstruction Finance Corporation through the Federal Reserve Bank of New York. They were commenced promptly. As in the case of the purchase of domestic gold, the policy was adopted of bidding for the gold substantially above the market price and of raising the bid prices at irregular intervals.

This plan was a unique one. Although stabilization funds used by central banks for the purchase and sale of gold and of foreign exchange have frequently been used to neutralize temporary disturbances in the exchanges and to assist in effecting a *de facto* stabilization, such funds have never, so far as I know, been used with the purpose of pushing up commodity prices to a much higher level intended to be permanent. Because of the unique character of the plan and of the absence of any explanation on the part of the administration as to just how the plan was expected to work, it was not understood by experts in the field of currency or of foreign exchange either in the United States or abroad.

Possible Explanations

Since there has never been an official explanation of the theory underlying the plan, the best that one can do is to make a guess as to what that theory was. Briefly stated, my own guess is as follows:

Prices were expected to be advanced by gold purchases chiefly in two ways:

First, it has been a common occurrence that when a paper currency was depreciating in value, the prices of gold and of the gold exchanges were among the first things to advance, and that wholesale commodity prices, and, later, the cost of living, moved upward in their trail. This, for example, was the experience of the United States from 1862 to 1865 during the greenback inflation [1] and of the principal countries having depreciated paper currencies during most of the time of the widespread inflation immediately following the World War. Many people, moreover, erroneously think of gold as a commodity possessing a fixed value which in some way is a true measure of changes in the value of all other classes of goods. What was more reasonable, therefore, on the basis of such assumptions, than to expect that, if you pushed up the paper dollar price of gold, the rise in

[1] See pp. 64–8.

this price would drag after it commodity prices ?

Second, pushing up the American price of gold and the gold exchanges might be expected to stimulate the export trade, especially in agricultural products, and thereby to raise the prices of American exports—a rise that might be communicated to other goods. If, for example, a shipment of raw cotton had been made from the United States to France on October 25, 1933, the day the gold purchase plan began to operate, at a cost to the French purchaser of 100,000 gold francs, with exchange at 5·87 cents to the franc, the gross yield to the American exporter would have been $5,870. For the week ending December 1, the mean cable exchange rate in New York on Paris was approximately 6·13 cents to the franc. Therefore, at that time the same shipment of cotton, if it were sold at the same price in France, again yielding 100,000 gold francs, would have yielded the American exporter the gross sum of $6,130, an increase of $260 resulting from the rise in the exchange, caused in this instance by a depreciation in the value of the American paper dollar, since there was no appreciable change in the value of gold in France during this period.

Exchange profits of this kind in terms of paper dollars tend to stimulate exports, unless and until they are eaten up by what is usually a laggardly rise in production costs at home. This same advance in exchange which stimulated exports, however, would cut into the profits of importers by reason of the fact that they would now have to pay approximately $6,130 to buy 100,000 francs' worth of goods, say, for example, silks, in France, whereas before they would have had to pay only about $5,870. This rise in exchange would retard imports and tend to put up the prices of all goods imported from France into the United States. The principle is well illustrated in the familiar fact of to-day that Americans are less and less able to travel in France (such travel being analogous to an American import—an import headed off as it were and consumed in France) by reason of the low exchange value of the dollar in France, while Frenchmen now find travel in the United States unusually cheap (such travel being analogous to an American export to France) by reason of the high exchange value of francs in terms of United States paper dollars.

A third possible method by which these gold purchases might under certain circumstances tend to raise commodity prices is through increasing the country's gold reserves and thereby providing a basis for further bank-note and deposit-currency expansion.

And, finally, such purchases might tend to push up commodity prices merely through the fact that they led directly to an expansion in the circulating media. If money or deposit currency were created either directly or indirectly for the purchase of this gold by the Government, the increase in the volume of the circulating medium thereby caused would tend to push up commodity prices. This, however, would be equally true if the expansion of the circulating media were effected for the purchase of wheat or cotton or any other commodity by the Government or for open-market purchases of United States securities by the Federal Reserve banks.

Although all of these four reasons for the gold purchases may seem plausible, an examination of the conditions under which the purchases were made will show that the last one alone is of any consequence as a force for pushing up commodity prices.

Gold is not Money but a Commodity in a Paper Money Standard Country

Under a paper money standard the commodity price level is a question of the relationship, on the one hand, of the volume of goods and services which are being exchanged and, on the other hand, of the volume of money and of deposit currency with which these goods and services are being bought and sold. Money and deposit currency must always be interpreted in terms of their respective velocities of circulation. Every price in the United States under the paper money standard of 1933 expressed a value relationship between a commodity and the paper dollar. Into it gold did not enter. In the history of money, it has often happened that the unused gold reserves " backing " an inconvertible paper money were worth much more at current exchange rates than the entire volume of paper money in circulation which they were supposed to support.[1] Reserves that are not used have little influence on the value of the paper money in circulation unless the prospects of an early resumption of convertibility are strong. Under a paper money standard, gold is little more than a commodity that is bought and sold like any other commodity.

England's Recent Experience

Between September, 1931, when England went off the gold standard, and the end of November, 1933, the price of an ounce of British standard gold increased 61 per cent., but the British index number of wholesale commodity

[1] See, for an example, pp. 86 and 87.

prices during those twenty-six months remained fairly stable and was only 5 per cent. higher on December 13, 1933, than it was when England gave up the gold standard in September, 1931. Although the internal value of the British paper pound has been fairly stable since September, 1931, from that date until early in 1933 the value of gold throughout the world was actually rising, as evidenced by declining price levels in gold standard countries. In this connection, *The Economist*, in its issue of October 28, 1933, page 799, said:

> British experience in the last two years is convincing testimony to the insensitiveness of internal prices to changes in the external value of the currency. And if this is true in Great Britain, where a third of the economic life of the country is in direct contact with the outside world, why should it be any less true in America, where the importance of international trade is very much smaller?

A Case of Barter

The " purchase " of domestic gold with short-time Treasury notes yielding interest at the rate of one quarter of 1 per cent. a year had no direct effect whatever on commodity prices. It was merely a case of bartering an interest-bearing obligation for gold. Money did not enter into the transaction. Of course, if these Treasury notes were used as collateral for bank loans the result would be an expansion of bank-note circulation or of deposit currency, but that would be true regardless of the purpose for which the Treasury notes were originally issued.

Obviously, if the Government enters the gold market for the "purchase" of gold in competition with other buyers, it will tend to push up the value of gold in terms of commodities, but that has no direct effect upon commodity prices at home, which are expressed in terms of inconvertible paper dollars. So far as the United States was concerned, the Government was merely pushing up the price of a single commodity, namely, gold, that was, on December 30, 1933, about 64 per cent. higher than it was ten months before, instead of pushing up the prices of goods in general which were lagging far behind the price of gold on the advance and which it was the Government's declared purpose to raise. Such "purchases" tended to increase the spread between the price of gold and the commodity price level. In competing for foreign gold, moreover, by more or less arbitrary price advances, the United States Government disturbed the international exchanges and aroused needless antagonism abroad, particularly on the part of the few countries which were still striving to maintain the gold standard.

Plan not Sound as a Stimulant to Export Trade

When a country is upon an inconvertible paper money standard, the principal influence that the value of gold exercises upon the price level is the indirect one which it exercises through the country's trade with gold standard countries.[1]

If America's purchases of foreign gold by pushing up its exchange rates on gold standard countries temporarily tended to stimulate exports, they equally tended to retard imports, and a country does not get rich nor effect a healthy advance in its price level by giving to foreigners more and more and receiving from them in return less and less. Furthermore, exports artificially stimulated by currency depreciation are likely to be artificially checked in the importing countries by anti-dumping laws, quotas, and tariffs raised to compensate for this currency depreciation. Other countries, moreover, can play the same game of trying to stimulate their exports by a continual process of currency inflation, which means international competition in inflation, with resulting

[1] See pp. 33–6.

chaos in the world's foreign trade, its currencies, and in all operations in international finance.

Gold Purchases to facilitate Currency Expansion by increasing Reserves

The argument that such gold purchases by increasing the country's gold reserves would make possible an expansion of bank-note and bank-deposit currency and, as a result, would bring about a rise in commodity prices, had no weight so long as the country's currency of the United States was on a paper money standard and not a gold standard. Furthermore, America's idle gold reserves were by far the largest gold reserves in any country in the world and averaged slightly larger for the year 1933 than for the boom years 1928 and 1929, when the country was doing about twice as much business annually in terms of dollar values as it did in 1933. A reduction in the gold content of the dollar, which at that time seemed to be contemplated by the Administration some time in the future, would, moreover, greatly increase the gold reserve ratios. The four billion gold dollars of the then existing legal standard of weight and fineness, for example, then owned by the Federal Reserve banks and the Government would become six billion dollars if the gold content of the dollar were reduced, say, by one third.[1]

Only as an Inflation Device did the Gold Purchase Plan tend to raise Commodity Prices

The only important way in which such gold purchases tended to push up commodity prices, aside from increasing the velocities of circulation by weakening the public's confidence in its currency and encouraging a flight from the dollar, was through serving as a device for bank-note and deposit-currency expansion. This was done chiefly through the fact that the Treasury notes paid out for the gold tended to flow into the banks and to become a basis for the expansion of circulating bank credit. If the American Government were determined further to inflate the currency, a renewal of the purchases of Government securities by the Federal Reserve banks in the open market would have been a much better device for expansion than the purchase of gold. It would have been more easily understood and, being a domestic affair, would have aroused much less antagonism abroad.

As *The Economist* said:[2]

> If the President is committed beyond recall to securing a rise of prices by monetary means, almost any method would be preferable to this.

[1] See pp. 186–9. [2] November 4, 1933, p. 850.

Commodity Prices fell after the Gold Purchase Plan went into Effect

The real test of the plan, however, must be found in its results. It was an avowedly " reflation " plan and its object was to raise commodity prices. Actually it did little more than bring great disturbance into the foreign exchanges, arouse antagonism abroad, and weaken confidence in the dollar everywhere. The first purchases of domestic gold were made at the price of $31·36 an ounce, which was 52 per cent. above the parity of the gold dollar and represented a gold value of 66 cents for the paper dollar. By the end of the year 1933 the Government had pushed up its price to $34·06 an ounce, which was 65 per cent. above the parity of the dollar and gave the paper dollar a value of 61 cents gold. Despite the depressing of the gold value of the paper dollar by about 8 per cent. between October 25, 1933, and the end of the year, commodity prices in general fell during that period. The general index number for commodity prices published weekly by *The New York Times Annalist,* for example, was about 2·6 per cent. lower on December 26, 1933, than on October 24, 1933, and the wholesale price index number of the United States Bureau of Labour Statistics, like the wholesale price index number for England, remained unchanged throughout the last four months of 1933.

The Economist's National Committee on Monetary Policy early in 1934 submitted to all the individual members of the American Economic Association a questionnaire on current monetary questions. One of the questions was:

> Do you favour the present (as in November–December) gold-buying policy of the United States Government?

Of the 789 " Yes " and " No " answers received, 628, or 80 per cent., were " No ".

Early in 1934 the gold purchase plan was quietly discontinued.

CHAPTER IV

INFLATION, DEFLATION, REFLATION

The Meaning of these three High-sounding Words—The United States has enough Money and Bank Deposits to put Business on the much-sought Price Level of 1926—Why this has not been Accomplished

THESE three high-sounding words—" inflation ", " deflation ", and " reflation "—are to-day in everyone's ears. What do they mean and to what extent, if any, is there inflation in the United States at the present time? It is the object of this chapter briefly to answer these two questions.

It is always dangerous to give definitions, and this is particularly true when the definitions have reference to terms of widespread popular usage in a politically controversial field. There are a number of different meanings in which the words constituting the title of this chapter may be correctly used. The definitions that follow give the meanings in which I shall use them. They are meanings in common usage among economists and other students of monetary problems. The terms are not limited in their application to inconvertible paper money currencies, but apply equally to currencies on the gold standard, the bimetallic standard, the silver standard, and any other standard with which the world has had experience.

Inflation

Inflation exists in a country whenever the supply of money and of bank deposits circulating through cheques, so-called " deposit currency ", increases, relatively to the demand for media of exchange, in such a way as to bring about a rise in the general price level. The general price level is a composite made up of an average of the index numbers of the prices of all goods and services that are bought and sold in the country. It covers both retail prices and wholesale prices, and should include, in addition to commodity prices, wages, service fees, and security prices, all items being weighted according to their relative importance as elements in the monetary demand.

The supply of money and of deposit currency must always be interpreted in terms of their respective velocities of circulation.[1] A permanent doubling of the velocities, for example, would double the supply of the media of exchange as truly as would a doubling of the volume in circulation. The demand consists

[1] See pp. 147–53.

of the volume of goods and services to be exchanged.

According to this definition, the United States had paper money inflation during the early years of the greenback period, from 1862 to 1865; it had gold standard inflation from 1896 to 1914, due chiefly to the large increase at that time in the world's production of gold; and it had bimetallic inflation from 1851 to 1855 due to the same cause. Mexico had silver inflation for most of the time from 1893 until it established the gold standard early in 1905. During the World War, although it was nominally on the gold standard, there was a decided inflation in the United States, due in large part to a great expansion of deposit currency. The mild inflation in the United States in the year 1928 was a gold standard inflation, due chiefly to a substantial increase in the velocities of bank deposit circulation.

Deflation

Deflation is the reverse of inflation. It occurs whenever the supply of money and deposit currency decreases, relatively to the demand for them, in such a way as to cause a decline in the general price level. Applying this definition, the United States had bimetallic deflation for most of the time from 1820 to 1832, paper money deflation from 1865 to 1878, and gold deflation from 1879 to about 1893. The great deflation through which the country passed from 1930 to February, 1933, was at a time when it was on the gold standard, and was due primarily to a collapse in business confidence and a breakdown of the machinery of circulating bank credit, with resulting substantial reductions in bank deposits and, proportionately, much greater reductions in the velocities at which the deposits circulated.

Reflation

" Reflation " is a word coined in the United States by the present economic depression. It is not found in the dictionary. It means to inflate (or deflate, as the case may be) the currency " back " so as to restore a pre-existing price level.

The price level usually contemplated when the term " reflation " is used in the United States is that of the year 1926, or, roughly speaking, of the period from the early summer of 1921 to October, 1929, which, as previously noted, was a long period of an unusually stable price level in the United States following the collapse of war-time inflation in the latter part of 1920 and early 1921.

" Reflation ", in the parlance of to-day, therefore, means an expansion in the supply of the circulating media, relative to the demand, to the extent necessary to restore the price level of the years immediately preceding the crisis of October, 1929; a price level which it is nowadays believed by many would be a normal one for present-day America. The Administration in Washington since March 4, 1933, has apparently been working for such reflation with the primary object of relieving the debtor classes of the extra burden placed upon them by the great rise in the value of the dollar (as expressed in the decline in the general price level) since 1929—a rise of about 25 per cent. between February, 1929, and February, 1933 (the month before the Administration took office).

Many of the economic recovery devices adopted by the Government have been motivated particularly by the desire to bring about this reflation. These have included the abandonment of the gold standard, liberal loans to farmers on the security of their crops, large Government expenditures for public works, direct purchases of farm products by the Government, the processing tax with its accompanying bonus to the farmers, a liberal loan policy on the part of the Federal Reserve system accompanied by extremely low discount rates, a large expansion of circulating bank credit through heavy purchases by the Federal Reserve banks of Government securities in the open market, the "gold purchase plan" inaugurated in October, 1933, and the silver purchase plans of 1933 and 1934. To what extent has there been an expansion in the supply of the circulating media relative to the demand since pre-depression days as a result of these and other inflationary forces?

Extent of the Expansion of Money and Deposit Currency in Relation to the Demand

Physical Volume of Business. Measured in tons, yards, bushels, or other units not involving price changes, the physical volumes of production and business in the United States for the latest month for which figures are available, as compared with those for the year 1926 and the month of February, 1933, were as follows, according to two well-known indices of business activity:

BUSINESS ACTIVITY

Name of Index	1926	Feb., 1933	Mar., 1934	Per Cent. Inc.
1—*N.Y. Times Annalist* Weekly Bus. Index	100	56	72	29
2—Standard Statistics Index Indus. Production	100	52	66	27

Roughly speaking, the physical volume of business in March, 1934, was something like 70 per cent. what it was in 1926 and there was an increase of about 28 per cent. from February, 1933, to the spring of 1934.

Volume of Money in Circulation. The total monetary circulation at the end of March, 1934, was 10 per cent. larger than for June 30, 1926, and it was about 18 per cent. lower than in February, 1933, the time of the heavy hoarding immediately preceding " the bank holiday ". On the other hand, the " stock of money ", namely, the money in circulation plus that held in the United States Treasury and by the Federal Reserve banks, increased from June 30, 1926, to the end of March, 1934, by 59 per cent., and from February, 1933, to March, 1934, by 31 per cent. The " stock " of monetary gold, moreover, increased by 73 per cent. from June 30, 1926, to March 31, 1934, and by 76 per cent. from February 28, 1933, to March 31, 1934. These increases in American " stocks " of money and of gold are explainable chiefly by the 41 per cent. reduction in the gold content of the standard dollar under the provisions of the " Gold Reserve Act of 1934 ".[1] It is these " new 59-cent dollars " with which we are now concerned, for it is in terms of them that

[1] See Chapter XII.

all prices in the United States are now expressed and that all the country's business is being conducted.

Since about 90 per cent. of the business in the United States is done by means of deposit currency, the figures for bank deposits in commercial banks are of particular importance in estimating the volume of the country's circulating media. Unfortunately, figures to a late date are not available for all banks. On the basis of such figures as are available, it is probably a conservative estimate to say that the net deposits of all commercial banks in the country are to-day something in the neighbourhood of 75 per cent. of what they were in 1926.

If the country is doing only about 70 per cent. of the physical volume of business that it was doing in 1926, and if for doing that business it has 10 per cent. more money in circulation and a " monetary stock " 59 per cent. greater, a stock of monetary gold 42 per cent. in excess of its entire monetary circulation, and if it has about 75 per cent. as large a volume of circulating bank deposits as it had in the year 1926, there would appear to be much more than enough circulating media to maintain the commodity price level as high as that of 1926.

Other Evidences of an Abundance of Circulating Media

There are many other evidences of an abundance of circulating media in the United States as a whole, although there are, of course, districts which have had many bank failures where there may be a real scarcity. The member banks of the country, in the spring of 1934, were carrying reserves above the legal requirements to the amount of approximately a billion six hundred million dollars, by far the largest excess reserves, up to that time, in the country's history, and being equivalent to about thirty-seven times the average for the boom year 1929. These excess reserves were well scattered over the country. For a long time member banks have been borrowing practically nothing from their Federal Reserve banks, although they have had enormous amounts of paper eligible for use in borrowing. The cash reserves of the Federal Reserve banks have been high and the discount rates have been the lowest in the country's history. The current rates of interest on the best grade short-time paper during the latter part of 1933 and early 1934 had been reduced almost to the vanishing-point.

Prices

In view of this superabundance of circulating media, one would naturally expect that the commodity price level would be higher than it was in 1926. The facts were just the contrary. Taking the year 1926 as 100, various significant price index numbers were as follows for February, 1933, the month preceding the breakdown of the gold standard, and for March, 1934:

Name of Index	Feb., 1933	Mar., 1934	Per Cent. Inc.
1—Bureau of Labour Statistics Index numbers:			
(*a*) Wholesale prices, 784 commodities	60	74	23
(*b*) Wholesale prices, farm products .	41	61	49
(*c*) Retail prices, food	57	68	19
2—General prices, Federal Res. Bank of N.Y.	73	79	9
3—Cost of living, Nat'l Ind. Conference Board Index	69	76	8
4—Price of gold franc exchange . .	100	168	68
5—*Annalist*, Sensitive Commodity Prices.	74	102	38

Although the value of gold, as measured by its purchasing power in gold standard countries, such as France, Holland, and Belgium, was very stable from February, 1933, to March, 1934, the American paper dollar depreciated to such an extent in terms of gold that the price of gold francs in terms of dollars rose 68 per cent.

Prices in the United States, as shown by the preceding table, by no means advanced from February, 1933, to March, 1934, sufficiently to register this depreciation in the dollar's gold value. As always in such cases, price advances occurred with varying degrees of lag for different kinds of commodities and services. In times of inflation it is only after considerable periods—periods that vary greatly for different groups of commodities and for different commodities in each group—that the slack is finally taken up. Although the general price level rose only 9 per cent. from February, 1933, to March, 1934, the prices of farm products in general rose 49 per cent., those of the *Annalist* group of commodities with sensitive prices, 38 per cent., and those of the 421 common stocks covered by the Standard Statistics Index about 79 per cent.

Money and Bank Deposits are circulating very Slowly

The real crux of the situation is not scarcity of circulating media, but the fact that money and bank deposits are circulating very slowly.[1] The average annual rate of circulation of the

[1] See chart, p. 149.

demand deposits of the country's leading cities, for example, was less than three-fifths as high in March, 1934, as it was in March, 1926, while for New York City alone it was only a little over two-fifths as high. There were sufficient money and bank deposits subject to check to do a much larger volume of business than was being done in the spring of 1934, at a general price level as high as that of 1926 and even higher, if they would only have waked up and moved again at their old-time velocities. The all-important reason why they were not moving more rapidly was that the people who owned them and upon whose initiative the country must depend for any strongly forward business movement were afraid to go ahead. They had little confidence in the business situation, and they were afraid of the future value of the dollar, in terms of which all contracts looking to the future must be made.

Inflation is like a Habit-forming Stimulant

For some time the United States has been injecting into its body economic the artificial stimulus of inflation and it is continuing to pump this habit-forming drug into the body. The body has shown a remarkable resistance, and so far the dosing has stimulated it only mildly. Despite the strong inflationary measures, however, industrial production as late as February, 1934, was not so near the pre-depression level as that of a number of other countries, such, for example, as gold standard France, and as the United Kingdom, Sweden and Germany.[1] Public confidence in the currency, moreover, has been surprisingly well maintained in the face of actual inflation and threats of further inflation. The great danger now to be guarded against is that the public may become increasingly fearful of the depreciation of its paper currency as a result of the inflationary forces now at work, and of the continued agitation in Congress for further inflation, particularly silver inflation, and that there may be an increasing flight from the dollar. If this should take place, the velocities at which money and bank deposits would circulate, instead of being subnormal as they are at the present time, might be increased manyfold. The more of the inflation stimulant the patient received, the more he would demand, and the discontinuance of the drug would be exceedingly painful both for the patient and for the doctor.

[1] See League of Nations, *Monthly Bulletin of Statistics*, April, 1934, p. 154.

CHAPTER V

THE GREENBACKS

The Previous Experience of the United States with a Paper Money Standard—The Story of Seventeen Years of "Temporary and Emergency" Inconvertible Paper Money and Their Lessons for the Present Controversy

"HISTORY teaches only one lesson, namely, that we learn nothing from history," is a cynical remark attributed to a prominent political scientist.

Every time one submits an economic question that has found its way into the field of political controversy to the test of historical experience either at home or abroad, the response is met, that "conditions are different to-day". And yet Plutarch was right: history does repeat itself. Human nature is much the same the world over and changes very little from generation to generation. One of the outstanding facts in the world's economic history is the way that history repeats itself in matters relating to paper currency, whether one is studying the experiences of ancient China, colonial America, France in the days of John Law or of the notorious assignats, or Europe during the World

War and immediately thereafter. This chapter and the next one will discuss two such experiences, in the belief that they contain valuable lessons for our present monetary controversies.

The depreciated United States note or " greenback ", as it was popularly called, dates from early 1862 to the end of 1878. During that entire time it was a subject of heated economic and political controversy and of strong sectional feeling. It broke both major political parties for a time into hostile factions, and created the Greenback Party, with a presidential candidate of its own in the election of 1868. To both historians and economists the greenback history is still a field of controversy.

A Wartime Temporary Emergency Measure

In 1860 the currency of the United States was on a metallic basis—legally the standard had been bimetallic since 1792—and consisted of gold coins, fractional coins of silver, nickel, and copper, and the notes of State banks. Out of a total monetary circulation in the loyal States at the beginning of the Civil War of between $255 and $280 millions the notes of State banks amounted to about $150 millions.

The first issue of $150,000,000 of greenbacks was authorized in February, 1862, more than a year after specie payments had actually been suspended and at a time when the bank-note currency was already depreciated in terms of gold about 3 per cent. The greenbacks were issued as a temporary emergency measure to assist a Government, largely dependent upon customs duties for its revenues, to meet the heavy fiscal demands suddenly thrown upon it by the outbreak of a great civil war. These legal tender notes, as they were often called, were issued with great reluctance on the part of the Lincoln administration and of most of the members of Congress who voted for them, and in face of a strong opposition both in Congress and throughout the country.

Shortly before favouring their issue, the Secretary of the Treasury, Salmon P. Chase, had held that great care would be necessary to prevent the issue from degenerating into an irredeemable paper currency " than which no more certainly fatal expedient for impoverishing the masses and discrediting the Government of any country can well be devised ". Senator Fessenden, chairman of the Finance Committee of the Senate, in the debate over the first legal tender bill, undoubtedly expressed the sentiment of most of the members of Congress who voted for the bill when he said : [1]

[1] *Congressional Globe*, February 12, 1862, p. 763.

> . . . Everybody who has spoken on this question, I believe, without an exception—there may have been one or two—but all the opinions that I have heard expressed agree in this: that only with extreme reluctance, only with fear and trembling as to the consequences, can we have recourse to a measure like this of making our paper a legal tender in the payment of debts.

Among the leaders of the opposition to the bill was Justin S. Morrill, Senator from Vermont, who dubbed it "a measure not blessed by one sound precedent, and damned by all", called it "immoral" and "a breach of public faith", and characterized those responsible for the bill as "quack doctors", administering "quack medicine to relieve a patient that is in no need of any medicine at all". He advocated increased taxation and a proper loan policy for meeting the emergency.[1]

In the House, Representative Crisfield of Maryland called attention to the enormous extent of existing debts, national, state, municipal, corporate and private, and the great volume of the country's bank deposits, all contracted on a gold and silver basis and by this legal tender bill made payable in a paper money

[1] *Congressional Globe*, February 4, 1862, pp. 629–33, and February 19, 1862, pp. 886–7.

that was certain to depreciate. He called the bill a plan of " gigantic confiscation ", and said that it would destroy private credit and make the value of the monetary unit depend upon the vicissitudes of war and of politics.

Of course, then, as now and always, there were a few eloquent, table-pounding advocates of paper money that expected miracles from this greenback issue. One Senator, for example, declared: [1]

> There is no probability that a currency based upon the resources of a great nation . . . will depreciate 50 per cent. or even 5 per cent. No such paper ever did depreciate, and none such, I venture to predict, ever will.

Increase in Greenback Issues

Although it was generally expected at the time when the first issue of $150,000,000 was made that this would be the only issue and that it would be in circulation only a short time, hardly had the first greenbacks come off the printing press before a second issue of $150,000,000 was demanded, and, with much less opposition than the first issue received, was authorized by Congress. A few months later there followed a third issue of $100,000,000, and

[1] Senator T. C. Howe, of Wisconsin, *Congressional Globe*, February 12, 1862, p. 763.

this, in turn, was subsequently increased to $150,000,000. The maximum amount authorized, therefore, was $450,000,000, and the maximum amount outstanding at any one time was $431,000,000. In addition, there were issued something like $15,000,000 of notes in fractional denominations varying from three cents to fifty cents, known as " shinplasters ", and required to meet the need for small change when the depreciating greenbacks, under the force of Gresham's law, drove out of circulation the nickel and copper coins. The State bank-note issues, many of which continued until after the end of the war, and the newly created issues of national bank notes were all convertible into greenbacks on demand, and their values therefore moved up and down with the fluctuating value of the greenback.

Fluctuations in the Gold Value of the Greenback, 1862–65

Throughout the greenback period, gold was bought and sold as a commodity. Customs duties were payable in gold and likewise the interest and, for the most part, the principal of the national debt. Regular dealings in gold began the fore part of January, 1862, on the New York Stock Exchange. A second and less fashionable market later developed into an exchange known as " The Gold Room ", while there were in the city two other gold markets of lesser importance. There are accordingly available for the entire period 1862 to 1878, inclusive, daily prices of gold in terms of greenbacks. With only one important interruption, the fore part of 1863, the price of gold moved up continually and rapidly until it reached its maximum of $2·85 of greenbacks for $1 gold on July 11, 1864. This price gave the greenback dollar a gold value of a fraction over 35 cents. A reference to the daily prices of gold will show that not until over two years after Congress had enacted the first greenback law, did the price of a gold dollar reach $1·64½, the equivalent of $34·01 for an ounce of fine gold, which was the Government's purchasing price for newly mined gold the fore part of December, 1933, only nine months after the discontinuance of the gold standard at the time of the bank holiday in early March.

Greenback Prices and Wages, 1862–65

The comprehensive price and wage studies of the greenback period by Wesley C. Mitchell [1] contain a mine of information concerning price and wage movements for the years 1862 to 1879.

[1] Mitchell, Wesley C., *A History of the Greenbacks*, and *Gold, Prices, and Wages Under the Greenback Standard.*

Mitchell's index numbers, which are the ones cited throughout this chapter, use the year 1860 as the base of 100. Wholesale prices in general lagged behind gold on the rise, and reached their actual maximum of 185 in the year 1865. Retail prices and the cost of living in turn lagged behind wholesale prices, reaching in 1866 their respective maxima of 180 and 170. Wages in their turn lagged behind the cost of living on the rise and did not overtake it until 1867, at an index number of 168.

Of course, the prices of different classes of commodities, as, for example, agricultural products, minerals, and manufactured products, and likewise of different commodities in each class, advanced very unequally; and the same was true of the wages of labour. Since the cost of living rose faster than money wages during the war, real wages declined and labourers suffered. This has been the experience in all countries and at all times with inconvertible paper money currencies; and it is such experiences in France of many generations ago that gave rise to the French saying: " The guillotine follows the paper money press; the two machines are complementary one to the other."

Farmers gained somewhat by reason of the fact that they sell at wholesale and buy at retail, for retail prices advanced less rapidly than wholesale prices. During these early greenback years, however, their gain from this fact was not great, for farm products in general did not rise as fast as the index of retail prices of all commodities prior to November, 1863. Broadly speaking, debtors gained at the expense of creditors, but since a large percentage of the people were both creditor and debtor, there was an offsetting of gains and losses. A farmer with a mortgage on his farm paid his interest and his principal—if he was actually making payments on the principal—through the sale of less bushels of wheat or of corn, or less pounds of beef or of pork, than he would have been required to had there been no inflation; and the labourer paid the mortgage on his home with less days of labour. Inasmuch, however, as the cost of living was also rising rapidly, it often happened that both farmer and labourer found that after meeting their increased expenses, they actually had left for paying debts less of the depreciated greenback dollars than they had had of gold standard dollars prior to the war. Moreover, if either farmer or labourer had a bank account or an insurance policy or owned any bond or mortgage himself, he saw its purchasing power declining rapidly.

Movement at End of War for Early Return to Gold Standard

When the war ended, the money of the United States, aside from the large issues of Confederate paper money which had become worthless, and of some gold coin that continued to circulate on the Pacific coast, consisted chiefly of about \$431,000,000 of greenbacks, \$15,000,000 of United States Government fractional notes, \$146,000,000 of national bank notes, and \$143,000,000 of notes of State banks, together with a small amount of minor coins. All this money circulated at parity with the greenbacks. The great currency problem before the country was therefore to bring this money back to parity with gold and silver.

The Secretary of the Treasury in December, 1865, said that the greenbacks " ought not to remain in force one day longer than shall be necessary to enable the people to prepare for a return to the constitutional currency ", and he favoured a vigorous policy of currency contraction, with the object of bringing the paper money back to gold parity as soon as possible. With this view both the public and Congress were in sympathy. On December 18, 1865, the House of Representatives, by a vote of 144 to 6, adopted a resolution of cordial concurrence

" in the view of the Secretary of the Treasury in relation to the necessity of the contraction of the currency with a view to as early a resumption of specie payments as the business interests of the country will permit ", and it pledged its co-operation to that end.

Increasing Opposition to Deflation

Enthusiasm for currency contraction, however, was short-lived, and after the early beginning, progress was exceedingly slow. Every forward step met strong opposition, and more than once the forces of inflation temporarily triumphed. In opposition to a contraction of the currency and a raising of the gold value of the greenbacks, practically all of the arguments now so frequently heard against currency deflation were advanced. Contraction, it was said, would prove burdensome and unjust to the debtor classes, would depress business, cause unemployment by bringing about declining prices, and would decrease Government revenues. For a period of thirteen years after the end of the war, a bitter fight was waged for and against an early return to a specie standard, culminating in the famous Resumption Act of 1875, which created the machinery by which the country was brought to the gold standard, January 1, 1879.

Price and Wage Movements, 1865–78

For these thirteen years, prices in general throughout the country tended downward. The annual average price of a gold dollar in terms of greenbacks declined sharply from $2·03 in 1864 to $1·12 in 1871. It remained at about that level until 1877, and then moved slowly to parity in 1879. Wholesale prices declined every year without interruption from 1865 to 1879, lagging behind gold until 1877 and then dropping even more rapidly. Retail prices and the cost of living declined almost continually from 1866 until 1879, but at a much slower rate than wholesale prices. Wages continued their wartime advance for six years after the war, reaching a maximum index number of 179 in 1871, and then started downward, thereafter declining every year until they reached 138 in 1879.

Thus, in general, the classes that gained during the years of wartime inflation were the ones that suffered during the long years of post-war deflation; while those who suffered from the inflation, gained by the post-war deflation; but this generalization requires many qualifications. Furthermore, the persons who belonged to one class during the inflation period often belonged to the other class during the period of deflation. Farmers, who usually sold at wholesale and bought at retail, suffered from the failure of the cost of living to fall as rapidly as the prices of their agricultural products. Labourers benefited through the facts that their wages continued to rise for some time after the cost of living began its long decline, and thereafter lagged behind the cost of living as it went down. But continually declining prices retarded business and as a result there was much unemployment. The year 1867 and the five years, 1873 to 1877, were years of business depression.[1] Creditors benefited at the expense of debtors, but collections were much more difficult when prices were falling, and there were many bankruptcies.

The Question of Paying Government Bonds in Greenbacks

In the post-war history of the greenbacks, a subject of much controversy was the question whether the principal of Government bonds must be paid in specie or could be paid in greenbacks. The organic law authorizing the issue of greenbacks made them

> receivable in payment of all taxes . . . due to the United States, except duties on imports, and of all claims and demands against

[1] Compare pp. 172 and 173.

> the United States of every kind whatsoever, except for interest upon bonds and notes, which shall be paid in coin,

and declared the greenbacks to be "a legal tender in payment of all debts, public and private, within the United States, except duties on imports and interest as aforesaid".

The interest on Government bonds and notes was clearly payable in coin, but how about the principal? This question was apparently not seriously considered in the early days of the war, because nearly everyone expected at that time that the war would be brief and that the depreciated paper money would be retired long before any appreciable payments on the principals of the war bonds would be required. While the laws authorizing some of the bond issues after 1861 provided that the bonds should be payable, principal as well as interest, in coin, many of the laws were silent as to the money in which the principal should be paid. When the bonds were issued, it seems to have been the general understanding—although this claim has been emphatically denied by some—that the principal as well as the interest would be paid in specie. This general belief undoubtedly made it possible for the Government to borrow on much better terms than it would have obtained had the expectation been that payment would be made in paper money. Concerning this controversy, the Secretary of the Treasury, Hugh McCulloch, in 1867, said:[1]

> . . . The bonds were negotiated with the definite understanding that they were payable in coin. . . . The contracts were made in good faith on both sides, a part of them when the Government was in imminent peril and needed money to preserve its existence, the balance when its necessities were scarcely less urgent. . . . Good faith and public honour, which to a nation are of priceless worth, require that these contracts should be complied with in the spirit in which they were made.

Nevertheless, in selling these bonds, the Government received depreciated paper money at par and such bonds as were not made by law specifically payable in coin as to their principal—and that meant most of the bonds—were, according to the strict letter of the law, legally payable in greenbacks, which the Government had expressly made legal tender for all debts, public as well as private, except those specifically mentioned in the law or the contract. This fact, together with the fact that the purchasers of all bonds issued after early 1862 had paid for them in greenbacks, was the argument of those who

[1] Report of Secretary of the Treasury, November 30, 1867, xxiii and xxiv.

favoured paying the principal in greenbacks in all cases, except where the law or the contract specifically called for payment in coin. It was urged that " the money of the ploughholders should be good enough for the bondholders ". President Johnson in his message of December, 1868, had favoured a discontinuance of the policy of making full payments of the principals of such bonds in gold. In reply, the Senate, by a vote of 43 to 6, passed a resolution saying that, " properly cherishing and upholding the good faith of the nation [we do hereby] utterly disapprove of and condemn the sentiment and proposition contained in . . . the late message of the President ". A similar resolution was passed by the House by a vote of 155 to 6. President Grant, in his inaugural address of March 4, 1869, said :

> To protect the national honour, every dollar of Government indebtedness should be paid in gold unless otherwise expressly stipulated in the contract.

This continued to be the policy of the National Government until the problem disappeared ten years later with the resumption of specie payments.

Chapter VI

GERMANY'S INFLATION

A Movement which brought the Monetary System to a Point where Germany's 1913 Mortgage Indebtedness of $10,000,000,000 could (in 1923) be Paid Off with One American Cent—A Gigantic Engine of Wealth Redistribution that worked Blindly

In Germany we have a case of an inflation that was held in control throughout a great war and for some time afterward, and then broke away completely and ran amuck.

When the war broke out, Germany was on the gold standard and all her currency was maintained at a parity with her monetary unit, the gold mark, which had a value equivalent to about 24 cents United States gold. Although there were several kinds of paper money in circulation, none of them was important, except the notes of the German central bank, the Reichsbank, which constituted about 2·6 billion marks out of Germany's total coin and paper money circulation of 6½ billion marks.

Wartime Inflation

Immediately upon the outbreak of the war Germany went off the gold standard and the

Government took vigorous measures to bring into the vaults of the Reichsbank as much of the country's gold coin and bullion as possible. A strong, patriotic appeal was made to the public to sell to the Reichsbank for paper money their gold jewellery and ornaments; and a memorial medal of iron was given to those who did so. It bore the inscription:

> Gold I gave to defend my country.
> Iron I took for a badge of honour.

Restrictions on the issue of notes by the Reichsbank were lightened and a new form of war emergency paper money, the Loan Bureau notes, was authorized to be issued under Reichsbank control. They were to be used chiefly in the granting of credit to the smaller business concerns.

Since Germany expected the war to be a short one and hoped to recoup her war expenditures from a defeated enemy, she adopted the policy of obtaining her war revenue chiefly through borrowing, rather than from increased taxation. Politically it was the more popular method, at least for a short time. Germany's wartime inflation, therefore, was merely another case of the old story of exploiting the monetary function of Government—namely, the function of providing the public with a good medium of exchange—

for the benefit of the fiscal function of providing Government with an adequate revenue.

Inflation Virus worked slowly

Under the influence of the war, Germany's national expenditures increased fivefold from the fiscal year in which the war began to the fiscal year 1918 in which it ended. During the same four years her total national debt increased sixfold. Her circulation of paper money (other than emergency money) increased from less than 3 milliard marks at the beginning of the war to 29 milliard at the end of November, 1918. During the war the amount of Government treasury and commercial bills discounted and of advances made by the Reichsbank increased from 1·3 milliard to 22·1 milliard marks. The demand deposits of the Reichsbank increased from 858 million marks to 10·7 milliard marks, and the country's gold reserve from 1·25 milliard marks to 2·38 milliard marks. From the calendar year 1914 to 1918 the price of gold in terms of marks, as measured by New York-Berlin exchange rates, rose only 41 per cent. Wholesale prices in Germany slightly more than doubled, and wages for every class of labour for which wage index numbers are available covering these years moved strongly downward.

Considering the extent of the expansion of

the country's circulating medium during the war, the advance in prices up to the time of the armistice was strikingly moderate. *It usually takes considerable time for the inflation virus to work, as has been recently experienced in the United States.*

Early Post-war Inflation

The post-war history of German inflation may be conveniently divided into two periods. The first extends from the armistice until April, 1921, when the Reparations Commission fixed Germany's indebtedness to the Allies at the stupendous figure of 132 milliard gold marks; and the second extends from that date until the stabilization plan near the end of 1923.

It is not surprising that Germany resorted to further inflation immediately after the war. The fiscal urge for inflation was strong. Badly defeated after a long and expensive war, Germany had no credit abroad. Business was chaotic at home and both the physical condition and the morale of the people were low. Millions of men were returning from the front and looking for jobs, and millions of workers had to be shifted from the industries of war to those of peace. There was widespread political and social discontent, and the fires of revolution were being fanned by Russian Communists. This was a difficult time to increase taxes greatly, and the sale of Government bonds to the German public on any considerable scale was out of the question. Most of the public could not buy bonds, and the few people who could would not with the prospect of further inflation still strong. The line of least resistance, perhaps the only line of action possible for adequately meeting the immediate situation, was further inflation.

From the fiscal year in which the armistice was signed to the fiscal year ending March 31, 1921, the expenditures of the National Government were multiplied about threefold, while from April 1, 1918, to March 31, 1921, the Government's total debt was increased 136 per cent. Its treasury bills discounted at the Reichsbank were multiplied fivefold, and the total circulation of paper money, exclusive of emergency notes, was multiplied fourfold. The Government's tax receipts (including customs) were increased from the fiscal year ending March 31, 1918, to the fiscal year ending March 31, 1921, from 3·1 milliard marks to 39·6 milliard marks—a noteworthy record under the circumstances.

Prices and Wages, 1918–21

Let us now consider what happened to prices and wages during the early post-armistice years.

Measuring the mark price of gold by exchange rates between Berlin and New York, we find that the price of gold rose continually throughout these four years and was over seventeen times as high in 1921 as in 1918. Wholesale commodity prices in general averaged about sixteen times as high for the calendar year 1921 as for 1918. Cost of living index numbers are not available for the war and early post-war years, but are available for the year preceding the war and for the time after February, 1920. They show the trend usually found in periods of inflation. The cost of living rose, but at a less rapid rate than wholesale prices.

The data concerning wages in Germany during these years are very meagre and relate only to industrial labourers in Government railway service—skilled and unskilled—miners and their helpers in the Ruhr region, printers and the high, middle, and lower classes of Government officials, respectively. Wage figures are available only for " real wages ", or wages in terms of purchasing power over the cost of living, rather than for money wages. All wage index numbers are based on the year 1913 as 100.

In every group wages rose immediately after the armistice and in every one except the higher two groups of Government officials, they continued to rise through the year 1919. In every group they declined sharply in 1920 and then rose in 1921. For all classes of labour covered by the figures, however, except unskilled labourers in the Government railway service, real wages were lower in 1921 than in 1913. For unskilled industrial labourers in the railway service, real wages were the same for the year 1921 as for the year 1913, and there was only a very slight decline in the real wages of the lowest grade of Government officials. From 1913 to 1921 the decline for the other groups was as follows:[1]

Skilled Railway Labour . .	25 per cent.
Miners and Mine Helpers . .	11 ,,
Printers	31 ,,
High-grade Government Officials .	60 ,,
Middle-grade Government Officials .	48 ,,

During this period labour in general was about as fully employed as it was immediately preceding the war.

Securities[2]

Short-time Government paper during these years was maintained at practical parity. While the long-time bonds of the National Government

[1] Statistisches Reichsamt, *Zahlen zur Geldentwertung in Deutschland, 1914 bis 1923*, pp. 40, 41 and 43.

[2] *Ibid.*, p. 5, and Statistisches Reichsamt, *Wirtschaft und Statistik, passim.*

declined somewhat, the decline was not great. German Government 4 per cents., for example, stood at 87 in January, 1922, taking the 1913 price as 100 ; they advanced to parity in September. Although the prices of these bonds were fairly well sustained, the value of the money in which they were payable was rapidly fading away. The owner of the bonds " kept the bottles, but the wine was rapidly leaking out ". Although the price of corporation stocks rose during these years, the rise was nothing like sufficient to compensate for the depreciation of the mark. For example, taking the year 1913 as 100, the index number of the prices of corporation stocks in January, 1921, was only 280, while that for the cost of living was 1,180, and that for the money wages of unskilled labour was 1,140.

Period of Runaway Inflation

In contrast with the preceding inflation which was more or less orderly and controlled, the inflation during the greater part of the years 1922 and 1923 was wild. It finally assumed proportions probably never before known in the world's history. The rates of depreciation shown by the American Continental paper currency of Revolutionary War days, or by the notorious assignats of France a few years later, were not in a class with the speed with which the mark of the latter part of 1923 slid down the toboggan. This was a period in which the economic and financial life of Germany was dominated by the pressure of her erstwhile enemies, to exact from her a preposterous amount of reparations, a pressure which reached its extreme form in the occupation of the Ruhr by French and Belgian troops in January, 1923. Here we rapidly move into astronomical figures.

The expenditures of the German Government rose from 145 milliard marks for the year ending March 31, 1921, to over 8 billion marks two years later; and then leaped to 49 trillion marks the following year. The taxes for the year ending March 31, 1924, totalling about 5 billion marks, faded into insignificance when compared with the $191\frac{6}{10}$ quadrillion marks of Government Treasury bills discounted and held by the Reichsbank on November 15, 1923.

Currency and Bank Credit Expansion, 1921–23

The currency expansion of this period, particularly after the summer of 1922, was phenomenal. From 252 milliard marks in August, 1922, it rose to 2 billion marks in January, 1923, to 28 trillion marks in September, 1923, and finally reached a maximum of 497 quadrillion marks at the end of the year, shortly after a plan had been adopted for stabilizing on a basis of 1 billion paper marks to the equivalent of one gold mark (24 cents United States gold). Bank deposits also showed a phenomenal increase at this time. The deposit balances of the large Berlin banks, with their numerous branches scattered throughout the country, increased from 116 milliard marks at the end of December, 1921, to $1\frac{6}{10}$ billion marks a year later, and to over 1 quintillion marks at the end of 1923. For the Reichsbank alone deposits of all kinds rose from $36\frac{5}{10}$ milliard marks at the end of 1921 to 531 milliard marks a year later, and to 548 quadrillion marks at the end of 1923.

Flight from the Mark

With such an expansion of money and of deposit currency, the value of the mark declined precipitously. Its rate of decline, in fact, was much faster than the rate of increase in the volume of the circulating medium itself. This was due to a strong flight from the mark, which was expressed in a tremendous increase in the velocities of circulation. No one wished to keep money any length of time, either in the form of a bank dcposit or in the form of cash in the pocket, for its value or purchasing power was fading with startling rapidity. Prices of merchandise in the shops were often marked up several times a day. In the latter days of the inflation there was literally a stampede of the public to get rid of their money. At times when the mark was depreciating most rapidly, people, in their desperate efforts to save something from the wreck, often bought goods they did not need. Commenting on this situation as it existed about the same time in Austria, de Bordes says: [1]

> On days when there was a sharp rise in prices there would be a run on the shops. Prices would then increase from hour to hour. . . . On such days one could witness "ludicrous scenes in which some old bachelor would be seen buying swaddling clothes, because the local shop had no other wares left for sale, or another would invest in four dozen toothbrushes simply to get rid of his money by some means or other".

The gold value of the mark, as measured by Berlin-New York exchange rates, dropped from slightly over half a cent in December, 1921, to $\frac{1}{100}$ of a cent a year later, and to 23 billionths of a cent at the end of 1923.

[1] De Bordes, J. van Walre, *The Austrian Crown*, p. 163.

Hyperinflation

At this time there occurred the phenomenon of " hyperinflation ",[1] frequently found in countries with rapidly depreciating paper currencies, where the value in terms of gold, or of goods at current prices, of the total volume of money in circulation becomes less and less as the amount of money in circulation increases. For example, the gold value of the 80,000 milliard in circulation in February, 1921, was $1,307,000,000 United States currency ; while that of the 497,000 quadrillion in circulation in December, 1923, was $114,000,000, or less than $\frac{1}{11}$ as much.[2]

Another fact of importance is that, although the gold reserves of the Reichsbank and of the Darlehnskassen tended downward during practically the entire three years 1921 to 1923, inclusive, and were less than half as much at the end of the period as at the beginning, these gold reserves at the end of December, 1922, were worth at current exchange rates in Berlin on

[1] For a good detailed account of Germany's experiences with hyperinflation, see Graham, Frank A., *Exchange, Prices, and Production in Hyper-Inflation : Germany, 1920–23.*

[2] See Statistisches Reichsamt, *Zahlen zur Geldentwertung in Deutschland, 1914 bis 1923,* pp. 45–7 and 53 ; also, *Annual Reports of Reichsbank, passim,* and Young, John Parke, *European Currency and Finance,* I, pp. 537 and 538.

New York nearly eight times as much as the total volume of paper money in circulation (exclusive of emergency currency) at the time, and about three times as much as this total volume of money in circulation plus all of the deposits of the Reichsbank and all of the deposits collectively of the large Berlin banks with their branches scattered throughout the country.

Prices and Wages

At the end of December, 1921, wholesale commodity prices averaged thirty-five times as high as they were shortly before the war. A year later they were nearly 1,500 times as high, and in December, 1923, they were 1,262 milliard times as high. During most of this time the cost of living rose less rapidly than wholesale prices, but by December, 1923, it had practically caught up, having been at that time 1,247 billion times as high as in 1913.

In early January, 1924, upon arriving in Berlin, I went to a cigar store to buy a cigar. I gave the dealer one rentenmark, worth 24 cents United States gold, in payment for the cigar costing one-half a rentenmark, and received my change, equivalent to 12 cents United States gold, in the form of the old marks. This "change" amounted to 500 milliard marks, which at the old gold parity would have been equivalent to 125 milliard dollars, over twenty times the number of dollars in circulation in the United States to-day. It must not be thought, however, as it so commonly is, that it required a truck to carry this amount of money. What I paid was a small-sized note carrying the words *Eine Rentenmark*, and what I received in change was another note of about the same size bearing the words, *Fünf Hundert Milliarden Mark.*

Wages of all kinds rose much less rapidly than the cost of living during these two years, with those for unskilled labour rising most rapidly and those for skilled labour, " the white-collared classes ", and for Government officials lagging far behind. As the rate of the mark's depreciation increased, it became more and more common for labourers to spend the greater part of their pay on the day on which they received it. They literally rushed to the shops to spend their money before its value faded away in their hands, often buying goods in bulk to cover their needs until the next pay day.

Various devices were adopted in the latter period of the inflation to protect the working classes, such as shortening the interval between pay days, reducing the period covered by wage agreements, and sliding scale wages based in part upon the movement of cost of living index numbers or of the dollar exchange rate.

There were sometimes supplementary wage payments in the form of family allowances. Such devices were helpful, but for labour in general they accomplished comparatively little in reducing the heavy losses incident to inflation. During most of the period of the inflation, labour, fortunately, was fully employed.

Inflation and Debts

One of the outstanding effects of the inflation was the extent to which it wiped out the debts of the German Government, German corporations, and the German people in general.

Debts were payable, principal and interest, in the depreciated mark. The extent of the loss to the creditor and of the gain to the debtor in each case, of course, depended upon the rate at which the mark depreciated between the date on which a loan was made and the date upon which it was paid.

In the case of long-time bonds and notes, life insurance policies, and similar obligations issued before the war and becoming due in 1923, at the time of the greatest inflation, the loss to the creditor or the beneficiary was the greatest, for the value of the principal paid at maturity was negligible.

In the latter days of the inflation, when it was feared that there might be a stabilization of the mark at a value higher than the prevailing one, the phenomenon was said to have occurred that John Witherspoon described as having taken place over a century and a half ago with the American Continental paper money; namely, that of "creditors running away from their debtors, and the debtors pursuing them in triumph, and paying them without mercy". The entire mortgage indebtedness of the German people, which was estimated to amount to about 40 billion marks, or 10 billion dollars in 1913, could have been paid off in November, 1923, with one American cent.

In the summer of 1925 there was enacted a complicated revalorization law, which provided for the restoration in gold values of certain small percentages of different kinds of long-time obligations, the value of which had been practically destroyed by the inflation. This law at best reached only a very small part of the creditor classes in Germany, which had suffered during the inflation period, and the percentages of compensation given to those persons whom it reached were at best exceedingly small.

Security Prices

Stocks rose rapidly during 1922 and 1923, while the prices of bonds in general showed little change. The rise in stocks, however, was nothing like as great as would naturally have been expected, considering the extent of the depreciation of the mark and the practical wiping out of corporate debts as a result of this depreciation.[1] Comparing December, 1923, with the year 1913, we find that the prices of stocks on the average increased 269 milliardfold, while the wages of higher Government officials, a class of people for whom the wage rise was exceptionally slow, increased 509 milliardfold.

German Government 4 per cent. bonds did not show any great change in price before the autumn of 1922. They were quoted at 87 in January, 1922, as previously noted, and at par in September. At the end of the year, however, rumours became common that in case of stabilization the Government was likely to show special consideration to the holders of these bonds. A strong demand for them developed both in Germany and abroad, and, for a few months, the price soared to extravagant figures. The rumour proved to be false and the price later collapsed. In general, high-grade bonds in Germany, governmental as well as corporate, maintained their prices fairly well during the inflation in terms of marks, but the value of the mark in which they were payable

[1] See pp. 166–70.

faded into nothingness. In other words, while their prices usually remained at or near par, the prices of everything else rose and rose enormously, leaving them far behind.

Inflation, an Engine of Wealth Redistribution that worked blindly

Probably no group suffered more from the inflation than what was often called Germany's "stalwart middle class". Of their sufferings Angell says: [1]

> The group which suffered most from the inflation . . . was the group which was . . . least able to defend itself: the middle classes among the town dwellers. . . . Composed largely of people with small fixed incomes, such as salaried officials and clerks, recipients of pensions, and little investors living on interest and rent—of whom the latter group were hit especially hard by the Government control of city rentals—they were precisely the group most exposed to the evil consequences of currency depreciation, while they lacked both the knowledge and the opportunity to combat it. Their savings disappeared [and] their pensions and annuities melted away. . . . Hundreds of thousands of educated men and women, too old or feeble or untrained to earn their own living, were abruptly faced with starvation. Many died. The others,

[1] Angell, James W., *The Recovery of Germany*, pp. 38, 39.

passing from day to day without hope, survived only by the sacrifice of treasured books, furniture, jewellery, and all their saleable possessions, and at the end by domestic and foreign charity. Their history is one of the most genuinely pitiful chapters in all the war and post-war tragedy.

Thus the inflation proved to be in Germany, as it has in every other country in which it has been practised on a large scale, a gigantic engine of wealth redistribution ; an engine that worked night and day and worked blindly, taking wealth here and giving it out there, robbing one economic group or one social class for the benefit of another, and doing it all without rhyme or reason, and with no regard whatever to the merit or demerit of those benefited and those despoiled.

Chapter VII

THE SILVER QUESTION

In the United States it is a Political Problem, not an Economic One—Seven States produce Silver to the Value of One-half the Peanut Crop, but these States have Fourteen Senators

SILVER, which in one form or another dominated the monetary controversies of the world during the last third of the nineteenth century, is again playing an important rôle on the American monetary stage. It came into prominence suddenly and, to most of the American public, in a startling manner through the action of the first session of the Seventy-third Congress.

Precipitous and Drastic Silver Legislation of 1933

On April 17 a proposition to remonetize silver, with free coinage at 16 to 1, was defeated in the United States Senate by a vote of only 43 to 33. Moreover, a considerable number of the 43 who voted against the measure were avowed silver advocates and werc, apparently, merely waiting for the Administration to show its hand before taking a stand favourable to silver. Three months earlier such a proposition for the free coinage of silver had been tabled in the Senate by the decisive vote of 56 to 18.

In May the silver provisions of the Thomas amendment to the Agricultural Adjustment Act became law. They authorized the President himself to fix a coinage ratio between gold and silver, and " to provide for the unlimited coinage of such gold and silver at the ratio so fixed ". This was bimetallism pure and simple, despite the fact that the amendment said that a gold dollar should be " the standard unit of value ". It was left entirely to the decision of the President whether a bimetallic standard should be adopted, and, if adopted, whether it should be national or international bimetallism.

Thus, with very little debate in Congress, and with almost no public discussion outside, full authority was suddenly given to one man to commit a nation of 123,000,000 people to a bimetallic standard at a time when no such standard had existed anywhere in the world for approximately sixty years. And this was done despite the facts : (1) that two generations ago nation after nation had deliberately discarded bimetallism after having given it a fair trial ; (2) that, after a quarter of a century of subsequent agitation throughout the world for a return to bimetallism—an agitation accompanied by a number of international conferences and by two heated political campaigns in the United States centring upon it—bimetallism was everywhere defeated and had almost been forgotten, both in the United States and Europe, for a third of a century; and (3) that the subject of bimetallism had not been an issue in the Presidential campaign of 1932 and had not even been mentioned in the platform of any national political party.

Arguments for International Bimetallism that carried weight during latter part of Nineteenth Century

Three arguments for bimetallism that carried weight in the controversy of the early nineties were:

(1) The world's gold production had been low since about the time of the discontinuance of bimetallism in 1873, while the demand for gold had been growing rapidly, largely as a result of the increasing adoption of the gold standard by different countries of the world. This increasing monetary demand for gold in the face of a supply that was growing very slowly raised the value of gold in terms of goods. In other words, it made it necessary

for the public to give more and more goods for a given quantity of gold, which is another way of saying that it caused the prices of commodities to decline. There was, in fact, a great and almost continuous decline in the price levels of gold standard countries from 1873 until the early nineties. This decline had depressed business and had imposed growing burdens upon the debtor classes, particularly upon farmers, who were being compelled to give up more and more bushels of wheat or of corn and more and more bales of cotton from year to year to meet their mortgage indebtedness, as the prices of their produce declined. The adoption of bimetallism, it was argued, would, by bringing silver to the assistance of gold as a standard money metal, increase the money supply and stop the fall in prices. It might even bring about a rise in prices. This would give the debtor classes a much desired relief from their increasing debt burdens.

(2) A second argument closely related to the first was that, since, historically speaking, the production of silver often was increasing at times when that of gold was decreasing or not increasing so rapidly, while at other times the production of gold was increasing when that of silver was decreasing or was not increasing so rapidly, the value of gold in terms of commodi-

ties often was tending upward at the very same time that the value of silver was moving downward, and vice versa. If the two metals were tied together in a bimetallic system, it was argued, these ups and downs would tend to compensate each other. The result would be that a standard consisting of the two metals tied together by a fixed bimetallic ratio, under which the weight of the monetary demand would be continually taken away from the metal which tended to be the dearer, and placed upon the metal which tended to be the cheaper, would be more stable in value than a standard made up of either one of them separately.[1] A common illustration was, that two drunken men walking arm-in-arm would be steadier than either one of them walking alone.

(3) A third argument was, that by tying up the value of gold to that of silver at a fixed mint ratio, the value of gold would be pre-

[1] This shifting of the monetary demand is the central idea of the so-called compensatory principle of bimetallism. Although this principle is the foundation of the theory of bimetallism, very little is heard of it in present-day discussions and comparatively few of the modern proponents of bimetallism appear to know anything about it. The principle is too complicated to be explained here. A good explanation by a bimetallist will be found in Francis A. Walker, *International Bimetallism*, pp. 94–105.

vented from fluctuating materially in terms of silver, and, likewise, the value of silver would be prevented from fluctuating materially in terms of gold. The foreign exchange rates, therefore, between gold standard countries and silver standard countries, of which there were many at that time, would be stabilized, as well as the rates between these monometallic countries and the bimetallic countries, and commerce between such countries would be freed from the exchange uncertainties and the speculative exchange risks that were then prevailing.

The Argument for Bimetallism much weaker To-day than Forty Years Ago

There was considerable weight in all three of these arguments for bimetallism when applied to the conditions prevailing in the world during the eighties and early nineties of the last century. None the less, they never carried sufficient weight to bring about an international agreement for a return to bimetallism, and there were very few economists of standing anywhere in the world who believed that bimetallism could be successfully maintained, except by international agreement among several important countries. Proposals, such as that of Bryan, for national bimetallism never had much scientific standing anywhere. Despite many vigorous efforts to obtain an international bimetallic agreement among leading countries [1] —efforts extending over an entire generation and strongly favoured by important interests in the leading countries of the world—no such agreement was ever obtained. At the present time there is practically no active interest in bimetallism in Europe or, in fact, in any other important section of the world except the American silver-producing States, and there does not appear to be the faintest prospect of obtaining an effective international agreement for the establishment of bimetallism. The United States could not possibly maintain bimetallism alone, and to undertake it would be foolhardy.

The three arguments mentioned above as carrying weight forty to fifty years ago have much less weight to-day than they had then.

The first of the bimetallic arguments mentioned above was, that by making silver a standard money metal in addition to gold, the country's money supply would be greatly increased and, as a result, commodity prices would be raised. This " scarcity of gold "

[1] For an account of the various international monetary conferences which considered the question of international bimetallism, see Russell, Henry B., *International Monetary Conferences, Their Purpose, Character, and Results.*

argument is very much weaker to-day than it was a generation ago. In 1896 commodity prices in gold standard countries had been declining almost continually for twenty-three years, while for the thirty-seven years since 1896 the general price level in the United States has been rising most of the time. In fact, it rose in twenty-four of these thirty-seven years, declined in six, and remained practically unchanged in seven. The low price level from which the United States is now suffering has lasted for only a brief period of time.

Then, again, instead of having a low volume of gold production in the world, as we had most of the time from 1873 to 1896, the world's gold production is high and has been advancing at rapid rates most of the time since the early nineties, except for a brief period during the World War and immediately thereafter. The world's gold production in 1897 was twice as large as the average for the five years 1886 to 1890. By 1915 the production of 1897 was doubled. For the nineteen years 1913 to 1932, the world's stock of monetary gold increased at an annual average rate of 4·8 per cent., while for the sixteen years 1913–14 to 1929–30, the average annual rate of increase in the physical volume of the world's production of basic commodities, according to the studies of Dr. Carl Snyder, was only 1·86 per cent. In 1931 the world's production of gold was larger than in any previous year in history, and in both 1932 and 1933 it was larger than in 1931.

Normally, India absorbs about 25 per cent. of all the gold produced each year. Recently she has been pouring gold on to the world's markets out of her hoards at an annual rate equivalent to about 45 per cent. of the world's total annual gold production. A larger proportion of the world's gold to-day is held in the so-called monetary uses, as contrasted with the merchandise uses, than for a long time in the past. Furthermore, the economies effected in the use of monetary gold during the last two decades, as a result of the growth of the use of bank cheques, increasing velocities of circulation, and the development of central banking, have been enormous.

There is no evidence whatever that the world is to-day suffering from an enduring shortage of gold. The present high value of gold in terms of commodities and the apparent scarcity of gold are due chiefly to the fact that the world, including governments, central banks, and individuals, is hoarding gold during the economic depression. A scared world is scrambling for gold and by far the largest supply of monetary gold that ever existed in the world's history is, for the most part, not working. The Fourth Annual Report of the Bank for International Settlements estimates the total amount of gold in private hoards at the beginning of 1934 to have been at 7,000 million Swiss francs, "or more than two and a half times the value of the current annual gold production". As soon as we find our way out of this slough of despond in which we are now grovelling, we shall discover that the world's supply of monetary gold is abundant [1] and that there is no need to supplement it by a resort to bimetallism for the purpose of preventing the price levels in gold standard countries from falling.

The "arm-in-arm drunken men" argument has lost much of its former weight in recent years, because for most of the time since 1896 the values of gold and silver in terms of commodities have been moving in the same direction. For example, although the value of gold was falling rapidly from 1896 to 1920, as evidenced by the great increase in commodity prices during those years, the value of silver in terms of commodities was falling even more rapidly during most of that time, so that an effective bimetallic standard during that period would presumably have been even less stable

[1] See pp. 178–81.

than our gold standard. The drunken men were staggering in the same direction.

A bimetallic tie-up with silver, therefore, would probably have dragged the value of gold down even faster than it actually declined, and would have given the country a more rapidly rising price level during most of the time from 1896 to 1920 than it actually had. From the middle of 1921 until the end of 1929 the value of gold was unusually stable—more stable than that of silver. Here, likewise, bimetallism would have done the United States more harm than good. During the latter part of the year 1920 and early 1921 and since January, 1930, an effective international bimetallism would probably have exercised a useful stabilizing influence, but these five years represent a very small part of the period that has elapsed since the defeat of Mr. Bryan in the election of 1896.

The third argument for stabilizing exchange rates between gold standard countries and silver standard countries is now of very little consequence. Except China, Hong Kong, and Ethiopia, no country in the world is now on the silver standard. Moreover, China herself has been contemplating the adoption of a gold standard as soon as conditions throughout the country become sufficiently stabilized to do so.

A stable exchange between China and the other countries of the world would be desirable, but China's foreign trade is a very small factor in world trade. In the prosperous year 1929, for example, China's total foreign trade constituted only about 2 per cent. of the world's total foreign trade. In that year China took only 2·1 per cent. of the total American exports, while America's imports from China were only 3·4 per cent. of its total imports.

" Doing Something for Silver " in Other Monetary Ways

If bimetallism, which to be effective would have to be international in scope, is economically undesirable and politically impossible under present-day conditions, why should not the United States use silver more extensively in other monetary ways ? In answering this question in the affirmative, the proponents of silver in this country frequently make two proposals. They are : (1) to increase the circulation of silver money ; and (2) to use substantial amounts of silver in addition to gold in the cash reserves of the twelve Federal Reserve banks, making it lawful for them to hold a certain proportion of their legal reserves in the form of silver. Let us consider each of these proposals.

Increasing the Circulation of Silver Money

Silver money in the United States takes three forms: fractional silver coins, silver dollars, and silver certificates.

American fractional silver coins of a half-dollar, quarter, and dime are now furnished to the public in such quantities as the public desires, and are acceptable by the Government and by the banks in all payments. If any considerable additional amount of these coins were put in circulation, they would be retired quickly through payments to the banks and by the banks to the Federal Reserve banks and the Government. The adjustment of the supply of these coins to the needs and desires of the public is to-day very nearly automatic.

Standard silver dollars have never circulated in the United States to any appreciable extent as standard money during its entire monetary history. A total of less than 11 million dollars were coined during the country's eighty-one years of legal bimetallism. As fiduciary coins they are heavy and inconvenient to handle and are everywhere unpopular. A number of efforts in the past on the part of the Government to force substantial quantities of them into circulation have dismally failed. The public prefers bills to " cartwheels ". The result is that silver dollars come back to the Government and to the banks almost as fast as they are put into circulation. For many years the proportion of the country's total monetary circulation consisting of silver dollars has been a declining one and to-day it is only about one-half of 1 per cent.

The silver dollar may be circulated by proxy through the silver certificate, which is backed dollar for dollar by silver dollars in the Government Treasury. In this form the United States has in circulation about $400,000,000 of silver. The silver held in reserve for these certificates, however, performs no useful function whatever. Before the country went off the gold standard in the spring of 1933, a dollar silver certificate always circulated as the equivalent of a dollar gold, and the Secretary of the Treasury was under legal obligation to maintain it at that value. The amount of silver in a silver dollar is worth in a free market only a small percentage of the money value of the dollar. For example, in the late spring of 1934 the average value of the silver content of the United States silver dollar in the free market of London, at current exchange rates, was equivalent to only 35 cents of the new American paper dollar.

The value of the country's silver coins does not vary at all with the wide fluctuations in the market value of silver. Moreover, it is a rare occurrence for the public to present silver certificates in any considerable quantities for redemption in silver dollars—the exchange is nearly always in the other direction—and the privilege of redeeming a dollar silver certificate in a silver dollar, worth, say 30 to 40 cents gold, has no influence whatever in maintaining the gold value of that certificate and will have no such influence unless the currency is inflated to such an extent as to bring the gold value of the certificate down to approximately the gold value of the silver in the silver dollar.

The investment of Government funds in the silver which is now tied up in the silver certificate reserves is a sheer waste, and sound monetary policy would counsel the substitution of Federal Reserve notes for these silver certificates, the breaking up of the silver dollars in the certificate reserves, and the gradual sale by the Government of the silver bullion thus obtained. This is essentially the policy that India adopted some years ago.[1]

[1] See pp. 123–5 and 127.

The Proposal that Part of the Legal Reserves of the Federal Reserve Banks should consist of Silver

Recently there has been agitation on the part of the silver people in favour of authorizing (or requiring) the Federal Reserve banks to hold part of their legal reserves in the form of silver, and a similar idea for the use of silver in central bank reserves has lately gained a little credence in Europe.

In harmony with this idea, President Roosevelt in a special message to Congress on May 22, 1934, recommended legislation authorizing "increasing the proportion of silver in the abundant metallic reserve back of our paper currency" and declaring it "to be the policy of the United States to increase the amount of silver in our monetary stocks with the ultimate objective of having and maintaining one-fourth of their monetary value in silver and three-fourths in gold".

Just what is meant by this latter clause is not clear. The term "monetary stocks" usually means all United States money in the country, including that in circulation, in the Federal Reserve banks and in the Treasury vaults. In the "Circulation Statement of United States Money", issued by the Treasury

Department as of March 31, 1934, the total amount or stock of money in the country was given, in round numbers, as 13·4 milliard dollars, or more than twice the amount of money in actual circulation. Of this sum 7·7 milliard dollars, or 57 per cent., consisted of gold; about 4·8 milliard dollars, or 36 per cent., consisted of various kinds of paper money (exclusive of gold and silver certificates); and 127 million dollars, or 1 per cent., consisted of minor coins of nickel and copper; while the total silver, amounting to 839 million dollars, or 6 per cent., was made up of 299 million dollars of fractional silver coins, 498 million dollars held in reserve against an equivalent in silver certificates, and about 52 million dollars additional in silver dollars. If the President's statement means what it says, then the stock of monetary silver in the United States now amounting to 839 million dollars would ultimately be increased to 3,363 million dollars; or, on the basis of the weight of our present silver dollar, would be increased by about 1,951 billion ounces of fine silver. This is equivalent to twelve times the world's annual production of silver in 1933, or to 91 times that of the United States.

To the contention that the proportion of silver in the nation's monetary reserves should be greatly increased, it may be asked: why in a gold standard country should the monetary reserves be held in silver dollars that are themselves liabilities [1] which will not circulate, or in silver bullion, the gold content of which is continually changing and which is not an acceptable medium to the public for redemption at home or for the making of international payments?

As we have seen, the public would not present their notes for redemption in silver dollars. They would not trust silver dollars for hoarding purposes and they could not use them for making payments abroad. Redemption in the form of silver bars, at continually varying market prices in terms of the gold dollar, would likewise be impracticable. The public would have little interest in silver bars, which would be clumsy to handle and the gold value of which would be continually changing. Some years ago, when silver was much more valuable than it is to-day, and when former President Taft was Governor-General of the Philippine Islands, his modest monthly salary, expressed in terms of the Mexican silver standard dollar, which was then circulating in the Islands, weighed approximately as much as he did. The custody of such a bulky article as silver in the

[1] See p. 122.

reserves of central banks would be expensive, and, likewise, its shipment in the making of international payments. If silver were to function like gold as a medium for making such payments, an agreement to that effect would be desirable with all the leading central banks of the world, and it is very doubtful if it would be possible to effect such an agreement. There is actually very little interest outside of the United States in the so-called " silver problem ", and the interest in silver in the United States is one that has been largely fomented by propaganda emanating from silver-producing states and from silver speculators. There has been no bimetallism in the world since 1873 and in no country, aside from the United States, is there any substantial advocacy of bimetallism. The " concerted action of all nations, or at least a large group of nations ", which the President is seeking, for greatly extending the use of silver as standard money is a political dream that has not the faintest chance of coming true.

The advocates of the plan to permit silver to constitute part of the legal reserves of the Federal Reserve banks seem to believe that the mere holding of this silver in the reserves would be sufficient. If the value is therc in the form of silver, even if the silver is never used, they reason, it will support or " back up " the value of the Federal Reserve notes and deposits against which it is held. This is a false and extremely dangerous idea of the principle on which a reserve fund functions.

A Reserve is a " Regulator Fund ", not a " Backer Fund "

A reserve that is not used is ineffective, and a reserve functions not as a backer fund but as a regulator fund. It was previously shown how at one time in 1922 during the German inflation, while the notes of the German Reichsbank were rapidly depreciating in value, the gold reserve of the Reichsbank was worth at current exchange rates eight times as much as the total amount of the bank's notes in circulation, and about three times as much as all the notes outstanding plus all the deposits of the Reichsbank and all the deposits of the large Berlin banks.[1] There are many examples in monetary history of such a situation, where a gold reserve held against an inconvertible paper money that was continuing to depreciate was actually worth much more than all of the paper money it was supposed to " back ".

The chief function of a central bank reserve is that of serving as a regulator fund to adjust

[1] See p. 86.

the supply of the currency to the changing demands of trade. When, under a normally functioning gold standard, the supply of money in the country becomes excessive relative to the demands of trade, as compared with the supply of money in other countries, money becomes cheap at home relative to the money abroad; prices of commodities and of securities tend upward at home as compared with such prices abroad; exchange rates move toward the gold export point; and, when that point is reached, gold becomes sufficiently more valuable abroad than at home to make its exportation profitable, and it is exported.

Paper money and underweight silver coins cannot take the place of gold for export because they have little or no recognition outside of the issuing country while gold is the money of international payments. These gold exports are continued until the exchange rates fall below the gold export point, and the currency supply is reduced to a quantity which brings the price level of the country back into equilibrium with the price levels of other countries. In other words, this process continues until the reduction of the country's currency supply has made the monetary unit so valuable at home that further exportation of gold becomes unprofitable.

Under conditions of currency redundancy like those assumed above, the central bank must be in position to give out gold freely for exportation to relieve the country of its relatively redundant currency, and to force exchange rates below the gold export point, thereby bringing the country's price level back into equilibrium with the price levels of the rest of the world. If the excess money is not permitted to flow off in this way, the country's currency is likely to depreciate.

The description given above of the fundamental principle underlying the normal functioning of a central bank's gold reserve suggests the principle that should determine the size and character of the gold reserve. Obviously, the normal reserve should be sufficient to provide for the absorption, through redemption in gold that may be exported, of any currency in circulation that may be rendered excessive or relatively redundant by the usual fluctuations in business. In addition, it should be large enough to afford a reasonable margin of safety for extraordinary business depressions or financial crises.

A central bank's reserve should therefore not be looked upon principally as a " backing " for the bank's note circulation and deposit liabilities, nor as a guaranty of the bank's ability to meet these obligations. All the assets of a central bank perform this function. A central bank's reserve is not concerned so much with the question of the ultimate solvency of the bank, as with the question of the machinery needed to maintain the gold value of the bank's notes and deposits through adjusting the supply of the circulating media—bank notes and deposits payable by cheque—to the changing demands of trade, increasing their circulation when trade demands increase and decreasing it when they fall off. In this way the currency at one time is prevented from depreciating because of relative redundancy and at another time from appreciating because of relative scarcity. The monetary supply is thus continually adjusted to the monetary demand. A central bank's reserve, therefore, is primarily a "regulator" fund, and it functions successfully only when it is actually used to decrease or increase the country's circulating media. It must be large enough to serve as a buffer fund to absorb the expansions and contractions of the currency that are required to meet the changing demands of business. A dead gold reserve—namely, one which is not actively used for redemption purposes—is, therefore, practically useless, and performs no important function in the maintenance of the gold standard.

Silver held in the reserves of the Federal Reserve banks would probably prove to be as dead as is the silver now held by the Government against its silver certificates, and as were the vast accumulations of silver in the United States Treasury that resulted from the ill-fated silver purchase laws of 1878 and 1890.

Economists Overwhelmingly Opposed to Increased Use of Silver as Basic Money

Apropos of the President's message of May 22 recommending broadening of the monetary base by inclusion of 25 per cent. silver, it is significant that, in response to a questionnaire recently sent to all the individual members of the American Economic Association, the leading association of American scientific economists, the " Yes " and " No " replies to two of the questions were as follows :

Question. Do you favour an increase in the silver base of our currency :

(*a*) by additional purchases of silver ?

(*b*) by bimetallism ?

Of the teachers of all grades 88 per cent. replied " No " to the first question, and over 97 per cent. replied " No " to the second question. Of all the individual members of the American Economic Association 85 per cent. replied

" No " to the first question, and 95 per cent. replied " No " to the second question.[1]

Silver in the United States is a Political Problem more than an Economic One

The silver money problem in the United States is fundamentally not an economic problem, but a political one. It is assuming importance to-day not by reason of any monetary need of the country, but because there are seven States that produce substantial quantities of silver, each of which has two United States Senators. These men look out for the interests of their constituents. If they did not, they would not long remain in Congress. Although these seven States in 1931 produced about 95 per cent. of our entire silver product, they had less than 3 per cent. of the American continental population, and in the year 1931—a year in which the value of their silver production was very much larger than in 1932 or 1933—their entire silver output was worth less than 9 million dollars, or was equal to about 1 per cent. of the value of the national wheat crop for that year, and about half that of the peanut crop.

[1] The details of this questionnaire are given in *The Independent Journal of Columbia University*, February 19, 1934.

Chapter VIII

THE SILVER PURCHASE PLAN

A Plan that "Does Something for Silver" by paying a Large Government Bonus to American Silver Producers over a Period of Four Years—The Silver purchased will perform no useful Monetary Function in the United States, but the Removal of the Competition of this American-produced Silver in the World's Silver Market will help India sell her Enormous Surplus of Monetary Silver at Better Prices and lessen the Depressing Influence of these Indian Sales on the Silver Market

On December 21, 1933, the President issued a silver purchase proclamation which provided for the purchase by the United States Government of at least 24,421,410 ounces of silver annually for a period of four years, this sum being approximately the silver production of the United States for the year 1932. Speaking of this proclamation in his message to Congress about two weeks later, the President said that he was using the authority granted him by Congress to purchase all American-produced silver. The price to be paid for the silver so purchased was $1.29 an ounce, which, if paid in the gold dollar of that date, would have represented the free coinage of all newly mined

American silver at the Bryan ratio of 16 to 1, initiated at a time when the actual market gold-silver ratio was approximately 75 to 1. But this silver was to be paid for in the new depreciated dollar, not in the old gold dollar,[1] and, while the silver producer was nominally to be paid $1.29 an ounce, there was a strange provision in the plan, intended apparently to do honour to the letter but not to the spirit of " 16 to 1 ", whereby " the Director of the Mint, with the voluntary consent of the owner [of the silver], was to deduct and retain of such silver so received 50 per cent. as seigniorage and for services performed by the Government of the United States relative to the coinage and delivery of silver dollars ".

With these provisions, the net price per ounce the Government will pay for the silver is 64½ cents paper, or about 38 cents of the old gold dollar, representing an actual gold-silver ratio of about 54 to 1. This net price of about 64½ cents an ounce was 50 per cent. above the market price of silver on the day on which the proclamation was made—a price which, under the influence of a market anticipating governmental action favourable to silver, had been advancing for some time. These silver purchases by the

[1] The coinage ratio in the United States, under the new " 59·06 cents gold dollar ", is no longer 16 to 1, but 27 to 1.

United States are to be made in pursuance of agreements entered upon at the time of the London Economic Conference of the summer of 1933.

What is the explanation of this action, calling for the expenditure of many millions of dollars of government funds for the payment of a bonus to the producers of silver in a time of great economic depression, when the Government is running an enormous budgetary deficit? To answer this question we must go back a few years.

Causes for Low Price of Silver

From 1928 to 1932 the gold price of silver had been declining. It averaged 58·2 cents an ounce for the year 1928, and 25 cents for December, 1932; this last price being the lowest average monthly price of silver in the world's history.

Many causes contributed to this decline, of which the three most important were:

(1) The increase in the value of gold itself due to the world scramble for gold during the period of world crisis and depression. This increase in the value of gold meant a decline in commodity prices in all gold standard countries, and the price of silver naturally went down with the prices of other commodities. Many important commodities, including copper, cotton, and

wheat, declined during those years even more than silver.[1]

(2) The fact that the world's production of silver did not fall materially from 1928 through 1931, as the price of silver declined. This was chiefly because about two-thirds of the world's silver is normally produced as a by-product of copper, lead, and zinc, and, therefore, as long as the prices of these principal products are maintained at a reasonably profitable level for the producers, this production will be kept up and their by-product, silver, will continue to be thrown on the market, even though its own price is low and declining. The fall in the world's production of silver in the last two years was probably due more to the low prices of copper, zinc, and lead than to the low price of silver itself, although this also was undoubtedly a factor.

(3) The third cause for the recent low and declining price of silver is a monetary one, and relates to important economies that governments of late have been realizing in connection with the monetary uses of silver. These economies take three forms:

(*a*) The widespread substitution of paper money for the higher denominations of silver

[1] See Kemmerer, E. W., *The Effect of Low Silver Testimony*. H. Resol., 72 Cong. First Session, p. 274.

coin, and the substitution of nickel, bronze, and copper coin for the lower denominations of silver. In recent years paper notes have been replacing the larger silver coins to a substantial extent in the monetary circulation of India, Germany, France, the Philippine Islands, and many other countries. On the other hand, although the first pure nickel coin was not struck until 1881, there are now at least twenty-four countries with pure nickel coins, and the encroachment of these coins of nickel and of coins of nickel and of copper combined upon the field of small fractional money formerly controlled by silver coins has been large. These paper notes and small coins of nickel, copper, and bronze are popular and are usually much cheaper than silver coins.

(*b*) The high price of silver prevailing during the World War and for a few years immediately thereafter made the silver bullion in the silver coins of many countries more valuable than the nominal gold value of the coins themselves. Figuratively speaking, for many countries there was " more than a dollar's worth of silver in a silver dollar or in two half-dollars or in four quarters ". Partly to protect their silver money from the melting-pot, and partly for reasons of economy, many countries, including Great Britain, Norway, Sweden, Holland, Germany,

Brazil, and Greece, greatly reduced the amount of pure silver in their coins. Great Britain, for example, reduced the pure silver content of her fractional money from 925 parts of silver in 1,000 to 500 parts in 1,000, the balance being copper. The debased coins are not so attractive as those of a higher fine-silver content, but they serve their purposes reasonably well and meet little opposition on the part of the public. They represent much-needed economies to these governments in times like these.

(*c*) The third, and probably the most important, monetary cause for the recent decline and low price of silver is found in the action of the Indian Government with reference to its silver reserve. Except during the period of the World War, and for a few years thereafter, India, up to September, 1931, had been on the gold standard since 1899. She has not been on the silver standard since 1893. Most of her money, however, has long consisted of silver rupees and their fractions, and of notes against which the reserves consisted of gold, silver coin and bullion, and securities. The Indian Royal Commission of 1926, in advising that India should go from the gold exchange standard to the gold bullion standard, recommended that she should greatly reduce the silver part of her reserve, and, in compensation, increase the gold part. The Commission rightly said that: [1]

> Silver reserves are ordinarily out of place in a gold standard system. . . . The desirability of eliminating rupee coin from the Issue Department is emphasized when its worth as a reserve asset is examined. From the point of view of the management of the currency by the Bank, the rupee coin is indistinguishable from the notes it issues, in spite of the fact that the rupee, appropriately described as a "note printed on silver", is made of a more costly material than the paper note. Both are tokens depending for their purchasing power upon the limitation of their issue, and both require to be secured by an adequate amount of gold and gold securities to assure their external value.

The Commission believed that, with the growing use of one-rupee notes which it was recommending in place of rupee silver coins, a substantial reduction in silver reserves would be desirable. It accordingly recommended a reduction during a transitional period of ten years of the Government's silver reserve holdings from 247 million fine ounces to 73 million. This recommendation, which was promptly adopted by the British Government, created a potential supply of silver to be thrown on the world's

[1] *Report*, pp. 32 and 63.

market during the next ten years, equivalent to about four-fifths of the world's silver production for a normal year. The Government's sale of this silver began in 1927, and by March, 1932, had totalled about 128 million fine ounces. Nevertheless, so much silver has flowed from the circulation into the currency reserves since 1927 that the total reserves have greatly increased instead of having been depleted, and they amounted to about 352 million ounces on December 15, 1933, or about 43 per cent. more than the amount held in 1927, when the plan to reduce them was inaugurated. This enormous accumulation of silver seeking a market obviously exercises a very depressing influence on the price of silver the world over.

Silver at the World Economic Conference

This, briefly, was the silver situation facing the world in July, 1933, when the Economic Conference met in London. The American delegation attended the Conference with instructions to get action, if possible, favourable to silver, and one of the most influential delegates was the eminent proponent of silver, Senator Key Pittman, of Nevada. The only resolution of any consequence passed by the Conference as a whole—and this apparently was not taken very seriously by most of the delegates—was the one relating to silver. This resolution was approved unanimously by the representatives of sixty-six states.

The principal provisions of the resolution were the following, which were recommended to all the Governments represented at the Conference:

> (*a*) That an agreement be sought between the chief silver-producing countries and those countries which are the largest holders or users of silver, with a view to mitigating fluctuations in the price of silver; and that the other nations not parties to such agreement should refrain from measures which could appreciably affect the silver market;
>
> (*b*) That Governments parties to this Conference shall refrain from new legislative measures which would involve further debasement of their silver coinage below a fineness of 800/1,000.
>
> (*c*) That they shall substitute silver coins for low-value paper currency in so far as the budgetary and local conditions of each country will permit. . . .

These resolutions were contingent upon the coming into force before April 1, 1934, of the agreement mentioned in item (*a*). This agreement, which was made outside of the Conference and is not part of its official proceedings, was signed July 22, 1933, by the representatives of Australia, Canada, China, India, Mexico, Peru,

Spain, and the United States. It is usually called the "supplementary agreement", and provides, among other things:

(1) That the government of India shall not dispose by sale of more than 140 million fine ounces of silver during a period of four years, commencing with January 1st, 1934, . . . [representing a rate of 35 million ounces a year].

(2) That the governments of Australia, Canada, the United States, Mexico, and Peru, during the existence of this agreement, shall not sell any silver, and shall also in the aggregate purchase, or otherwise arrange for withdrawing from the market, 35 million fine ounces of silver [namely, the maximum amount which India is permitted annually to throw on the world's market] from the mine production of such countries in each calendar year for a period of four years, commencing with the calendar year 1934. The said Governments undertake to settle by agreement the share in the said 35 million fine ounces which each of them shall purchase or cause to be withdrawn.

(3) That the silver purchased or withdrawn in accordance with Article 2 above shall be used for currency purposes (either for coinage or for currency reserves), or be otherwise retained from sale during said period of four years.

(4) That the Government of China shall not sell silver resulting from demonetized coins

for a period of four calendar years, commencing January 1st, 1934.[1]

(5) That the Government of Spain shall not dispose by sale of more than 20 million fine ounces of silver during a period of four years, commencing January 1st, 1934. . . .

Passing over items 6 and 7 of the agreement, which are not important for our purposes here, we may note that the final item provides:

That this memorandum of agreement is subject to ratification by the Governments concerned. The instruments of ratification shall be deposited not later than April 1st, 1934, with the Government of the United States. . . . The Government of the United States is requested to take such steps as may be necessary for the purpose of the conclusion of this agreement.

President Roosevelt, in announcing his proclamation of December 21, 1933, said:

. . . I have to-day, by proclamation, proceeded to ratify the Treasury Department agreement with regard to silver, which has already been put into effect by the Government of India and which I understand other nations concerned are about to act on.

[1] Since China is on a silver standard and all her silver coins are full-weight coins, there would seem to be little danger from this contingency.

The other nations have now all ratified the agreement and are now in position to see how it will affect the different parties concerned.

The Silver Agreement and "the other 58 States"

Leaving out of account the eight nations that entered into the supplementary agreement, we have left fifty-eight whose delegates signed the general silver agreement of the Conference. How will these fifty-eight nations be affected by the agreement? Not one of them was on the silver standard, except Ethiopia, and not one produces any considerable amount of silver. Most of them produce no silver at all. Some, like Belgium, Denmark, Norway, and Finland, had no silver coins. Many of them have already substituted paper money of low denominations for their larger silver coins, and have substituted nickel or bronze coins extensively for their former silver coins of the lower denominations. Many of them in recent years have greatly reduced the percentage of silver contained in such silver coins as they have retained in circulation. In this group are to be found about thirty countries, including Austria, France, Germany, Great Britain, Hungary, and Poland.

Probably most of the countries that would be

interested in such economies in the use of silver as those contemplated by the resolution have already realized them. Certainly, few of these fifty-eight countries are buying silver for coinage purposes during these years of paper money inflation and business inactivity, while a number of them have surplus stocks of silver resulting from the coinage measures just mentioned for which they will welcome a market that is artificially buoyed up by the heavy silver purchases the United States has agreed to make.

India and Spain

The second group consists of India and Spain, which have large quantities of surplus silver that they wish to sell at the most favourable terms possible. Here we shall limit our discussion to India, which is by far the more important. India, as previously explained, has been for about seven years making efforts to sell her large surplus of silver, but, despite substantial sales, now has a much larger surplus of monetary silver than she had when she began. The heavy purchases of silver which the United States is now undertaking at artificially high prices and the withdrawal of this silver from the world's markets will prove a godsend to India, for they will assure her a better market and higher prices than she otherwise would have had. According to the agreement, India is authorized to unload on the world's markets annually the full amount of 35,000,000 ounces of silver, the equivalent of which the other nations that are parties to the special agreement are expected to absorb.

China

China is in a class by herself, so far as this silver agreement is concerned, for she is a silver standard country. The " whereas " of the supplementary agreement relating to China merely declares that " it is to the advantage of China that sales from monetary stocks of silver be offset by purchases as herein provided, with a view to its effective stabilization ". While the resolution of the Conference thus speaks of effective stabilization for China, most of the proponents of silver in the United States have been maintaining that China was suffering from the present low value of silver, and that a substantial advance in the price of silver would greatly benefit China " by raising the purchasing power of 400 million people ". The President's Silver Proclamation of December 21, 1933, specifically refers to the propriety of co-operating with other governments " to augment the purchasing power of peoples in silver-using countries", though just what is meant by silver-using countries is not clear. The only silver standard countries in the world to-day are China, Hong Kong, and Ethiopia.

Exactly how these heavy silver purchases by the United States are to help China is not evident. China, like every other country, pays for her imports chiefly by her commodity exports. Since nearly every year she has a heavy net importation of silver, as a merchandise proposition she would not find it to her advantage to have the value of silver go up and to be compelled to give more and more of her goods in exchange for a given amount of silver. As a currency matter, she would not gain from a large advance in the value of silver. China's commodity price level has been much more stable since 1929 than the price level in any gold standard country. Since January, 1933, however, commodity prices in China have been declining. The wholesale price index number for Shanghai, for example, fell every month but two from January, 1933, to February, 1934, and in the latter month stood at about 10 per cent. below the level in January, 1933. It is hard to believe that the proponents of silver in the United States, who have shown such great solicitude for the poor people of China and wept so many crocodile tears in their behalf, should want to adopt a policy that, by raising the value of silver in China and depressing commodity prices, would increase the debt burdens of the Chinese people and, in short, impose upon China all the hardships of deflation from which the silver group propose to relieve the American people by the remonetization of silver. Such a policy might well force China to an early adaption of the gold standard. If a rise in the value of silver in China should stimulate China's commodity imports, it would correspondingly retard her exports. This fact, along with the resulting lower price level in China, would tend to change China from a heavy importer of silver to an exporter of silver, with a resulting increase in the supply of silver being thrown on to the American, European, and Indian markets.

It is a significant fact that China's foreign trade since the crisis of 1929 has suffered less than that of the United States or of many other gold standard countries. In fact, China has maintained well her foreign trade position among the different countries of the world. From 1929 to 1932, for example, China's percentage of the world's total foreign trade declined from 2·13 to 1·90, while that of the United States declined from 13·83 to 10·92.

The Silver-producing States

The fourth group of countries to consider is that consisting of the leading producers of silver, namely, the United States, Canada, Australia, Peru and Mexico. Even if each of these latter four countries takes off the market its fair share of the silver it produces during the next four years, the United States would still be buying each year about five-eighths of the amount of silver that India and Spain were unloading on the market. Their Government would be taxing the American people in one form or another millions of dollars a year in order to pay their silver-producing interests a large bonus on their product, and to help India and Spain to sell at good prices their surplus silver. This bonus would greatly stimulate silver production in the United States and the Government would find itself under heavy pressure to continue to buy the entire American product.

Meanwhile, the United States would be piling continually higher its supply of monetary silver, which would not circulate to any considerable extent in the form of silver dollars [1] and which would be useless as a reserve for a gold standard country.[2] This vast supply of dead silver in

[1] See pp. 19 and 108.

[2] See pp. 19, 20, 21, and 108–10.

the United States Treasury would hang over the world's silver market like the sword of Damocles, in the way that India's great surplus has been doing recently. And when in the future the people of the United States should come to their senses on this whole silver question, they would want to sell it as India and Spain are selling their supply to-day. Where in that day will they find the generous Grandpa Nation that will do for them what they are now doing for India and Spain? Is it any wonder that the other sixty-five countries represented at the World Economic Conference, which could not agree on anything else of consequence, should have accepted enthusiastically the silver purchase proposition submitted by the United States?

Chapter IX

THE COMMODITY DOLLAR

This Plan provides that the Weight of the Gold Content of the Dollar shall be changed from Time to Time in Order that the Dollar shall always have the same Purchasing Power

THE term "commodity dollar" in the sense in which it is usually employed to-day is comparatively new, although the general idea which it connotes has been suggested from time to time over a period covering more than a hundred years.

The American astronomer, Simon Newcomb, suggested a commodity dollar plan in 1879, and an Englishman, Aneurin Williams, worked out such a plan in considerable detail twelve years later. Little attention, however, was given to these early proposals. Irving Fisher formulated a commodity dollar plan independently in 1911 and he, more than anyone else, is responsible for the subsequent development of the plan and for arousing public interest in it. Although a number of varieties of the commodity dollar plan appear in present-day discussions, the plan as developed by Fisher in his book on *Stabilizing the Dollar* is the one that is best known. It is the plan contemplated in this chapter. The formulation of the plan runs briefly as follows:

A Plan to give the Dollar a Fixed Value

Our present unit of weight is a fixed weight, the pound; our unit of length is a fixed length, the foot; and our unit of content is a fixed content, the quart. But for its unit of value, under the gold standard, the United States used not a fixed value, but whatever value happened to attach at the time to a fixed weight of gold, namely, to 23·22 grains of pure gold, which was the gold content of the dollar. Since the value of gold is continually changing under the impact of world forces of demand and supply, the country's gold unit of value was not fixed but variable, and this instability in the value of its monetary unit was a serious defect, especially in its bearing upon the relations between debtors and creditors.[1]

To meet this difficulty the commodity dollar plan provides that the weight of the dollar shall be changed from time to time with the object of making the dollar always buy approximately the same quantity of goods; in other words, of giving it a fixed value or purchasing power.

[1] See pp. 9–11.

To the minds of the advocates of the commodity dollar it is a misnomer to call it a "rubber dollar". In fact, their object is to transfer "the rubber" from the value of the dollar, which is its most important quality, to the weight, which in itself is unimportant.

The commodity dollar at the beginning would be made equivalent to a certain weight of gold. It would not be coined, but would circulate in the form of notes which would be convertible on demand into varying amounts of gold bullion known as "the bullion dollar".

How the Value of the Commodity Dollar would be Determined

The Government would maintain a price index number covering the principal commodities in the economic life of the country. At stated regular intervals, perhaps every two months, this index number would be published. When it rose, the weight of the bullion dollar would be increased, and when it fell, the weight of the bullion dollar would be decreased. For example, if the plan were inaugurated with a bullion dollar of 23·22 grains of gold, and if at the end of the first two months the price index number was found to have risen from 100 to 101, so that it would then require $101 to buy the same composite group of commodities that could have been bought by $100 two months before, the bullion dollar would be increased in weight, say, by 1 per cent., or from 23·22 grains of pure gold to 23·452 grains, and all bullion dollar notes would now become convertible into the bullion dollar of this increased weight. If at the end of the next two months the commodity price index number showed another advance, the weight of the bullion dollar would again be increased and this process would be continued every two months until the advance in commodity prices was checked and the price index number was forced back to 100. On the other hand, whenever the commodity price index at the end of any two months period was found to be below 100, the weight of the bullion dollar would be decreased, and a reduction would be made every two months until the index number was forced back to 100.

" How can we know ", asks Fisher, " that if the index number is 1 per cent. above par, a 1 per cent. increase in the weight of the gold dollar will be just sufficient to drive the index number back to par ? The answer is we do not know, any more than we know when the steering-wheel of an automobile is turned, that it will prove to have been turned just enough and not too much." If the first correction is not enough or if it is too much, the index number when next computed will tell the story, and the necessary corrections will be continued every two months until they accomplish their purpose.

The Government would purchase gold bullion, paying for it by bullion dollar notes at its reserve fund offices at the rate of one bullion dollar note for each bullion dollar, and, on the other hand, it would sell bullion dollars on demand for bullion dollar notes at a price slightly higher than a one-dollar note for one bullion dollar. This difference between the Government's buying rate for bullion dollars and its selling rate, which it is proposed at the beginning should be 1 per cent., is intended to prevent dangerous speculation in gold in anticipation of a rise or fall in the gold content of the bullion dollar. The weight of the bullion dollar would not be raised or lowered more than 1 per cent. at any one time, and this 1 per cent. charge, Fisher maintains, would eat up all the profits a speculator might otherwise expect to make by buying gold from the Government just before an anticipated reduction in the size of the bullion dollar and selling it back after the reduction ; or by selling gold to the Government just before an anticipated rise in the bullion dollar and buying it back after the rise. Fisher believes that the expenses and risks involved in any extensive speculation in bullion dollars for a longer period than two months would not be justified by the probable gain.

Criticisms of the Plan

Since Fisher first put forth his plan in 1911, there has been a large amount of controversy over it. It has been vigorously assailed by economists in the United States and in Europe and has been ably defended by Fisher and by a number of other economists. The plan has been modified from time to time to meet objections that have been raised. A detailed and critical presentation of the objections to the plan is impossible within the limits of a short chapter. I shall confine myself, therefore, to a brief summary of what I believe to be the principal valid objections to the plan.

A Long Period of Rising Prices would Dangerously deplete the Gold Reserve

The ratio of the bullion dollar reserve to the volume of notes in circulation would continually decline during periods when the commodity price level was tending upward, for, as the weight of the bullion dollar was increased, the number of bullion dollars represented by the gold reserve existing at the time the increases were made would decline. Experience has shown that rising prices may continue over long periods of time. Furthermore, the price advances that would need to be checked might often be very pronounced from year to year, or even from month to month. The wholesale price level in the United States, for example, rose almost continually from 1896 to the summer of 1920, reaching a maximum in May, 1920, of 260 per cent. above that of the year 1896. Within that period there were at least three months in which the monthly rise was 6 per cent. or over and at least six months in which it was over 4 per cent. Of course, the opposite would take place when commodity prices were falling and the weight of the bullion dollar was being progressively reduced. Then the percentage reserve would be increased.

Speculation in Gold would tend to Defeat Plan

A second objection that has been frequently urged is that, at times when the commodity price level was tending strongly upward or downward, the plan would give rise to heavy speculation in the bullion dollar that would render impossible the price adjustments sought. The proposed difference of 1 per cent. between the Government's buying- and selling-price for the bullion dollar and the restriction of changes in the price to 1 per cent. every two months were incorporated in the plan to meet this difficulty, but it is very doubtful if they would prove effective at times when the commodity price level was tending strongly in one direction or the other over substantial periods. This difficulty is discussed at some length by Dr. B. M. Anderson in his article, "On the Practical Impossibility of a Commodity Dollar", published in the *Chase Economic Bulletin* of December 13, 1933, from which the following quotation shows the general line of reasoning:

> We take first the case of falling prices. We shall assume that prices have fallen 2 or 3 per cent. and that the tendency is still downward, so that we can confidently expect that for the next two or three months the redeeming authority will progressively lighten the gold content. Will not foreigners withdraw their money from our markets, turning their cash in American banks into gold in order to avoid the loss of 2 or 3 per cent. which they can clearly anticipate? Will not speculators rush to turn in their dollars for gold, anticipating that at a later time they can turn back the same gold to the redeeming authority and get more dollars for it? Will not everyone who has foreign payments to make hasten to purchase foreign exchange,

thereby increasing the foreign drain upon our gold? And will not all of these transactions operate to reduce the money supply of the country—since it is by turning in paper money that the gold is obtained from the central authority? And will not all these operations withdraw money from bank reserves, tightening the money market, raising rates of interest on short loans, reducing the credit available for the carrying of commodities and securities? But is tightening the money market the correct procedure when one wishes to raise prices?

Commodity Dollar would bring Great Instability into our Foreign Trade

If the plan were a national one, in effect only in the United States, its operation would bring a large element of uncertainty and speculation in all of our foreign trade. Fluctuations in exchange rates between the United States and other countries would be wide. From the standpoint of foreign commerce, it would be in much the position of a country like China a few years ago, which was on the silver standard when most of the rest of the world with which it carried on trade was on a gold standard. This difficulty would be largely overcome if the plan were adopted, on the basis of an international price index number, by all the leading commercial nations of the world, but the prospect of such an international adoption is exceedingly remote. Twenty-one years ago, in answering this and other criticisms of the plan, Fisher said: [1]

> I am sure I am under no illusions as to the possibility of the early adoption of any plan to standardize the dollar. This may require centuries . . .

Powerful Psychological Factors that affect Price Movements are ignored by Plan

Perhaps the most weighty objection to the plan is that it involves a highly mechanistic theory of the relation of money and bank credit to prices, and in doing so underestimates the great influence of changing human emotions, feelings and prejudices, hopes and fears on the movements of our commodity price level, particularly in periods of economic and political stress and strain.

In the fourth chapter it was pointed out that the price level was the resultant of the interaction of the supply of circulating media, consisting of money and of bank deposits subject to cheque, and of the demand for these circu-

[1] *The American Economic Review, Supplement,* March, 1913, p. 46.

lating media represented by the physical volume of goods and services to be exchanged, and it was noted that the supply of the circulating media was not only a question of the volume of money and of bank deposits, but also of the velocities at which this money and these bank deposits circulate. In the days of the Bryan free silver campaign, the concept of the velocity of monetary circulation was often explained by means of the " ten clown illustration ". There are two scenes. In the first, ten clowns are lined up on the stage. Each clown owes the other a dollar and each clown has one dollar. At the tap of a bell, clown No. 1 hands his dollar to clown No. 2; clown No. 2 hands his dollar to clown No. 3, and so on around until clown No. 10 hands his dollar to clown No. 1. Here ten dollars' worth of money work has been performed by ten dollars and each dollar has circulated once. In the second scene the same clowns are on the platform and each clown owes the other a dollar. All clowns are broke, except clown No. 1. At the tap of the bell, he hands his dollar to clown No. 2; No. 2 passes it on to No. 3, and so on around until No. 10 pays the dollar back to clown No. 1. Here again ten dollars' worth of money work has been performed, but in this case it has all been done by one dollar, which has circulated ten times.

Sometimes money circulates very rapidly and at other times very sluggishly. Hoarded money has a velocity of zero.

In the United States, as previously noted, something like 90 per cent. of the total business is performed through the circulation of bank deposits by means of cheques, so-called " deposit currency ". Suppose a merchant, whose receipts and payments are practically all made by means of cheques, deposits daily in the bank all the cheques he receives and makes all his payments by means of cheques drawn on his own account. If this merchant keeps an average daily deposit balance at his bank throughout the year, of, say, $10,000, and if the total amount of cheques which he draws against the account in the course of the year is $100,000 his deposit has turned over ten times, or has an annual velocity of 10. If the amount of cheques charged to his account during the year amount to $200,000, the velocity has been 20, and if it amounts to $300,000, the velocity has been 30. Whether the velocity has been 10, 20, 30, or any other figure, the legal cash or deposited reserve which the bank has been required to hold has been the same, for that is a fixed percentage of the deposit balance and does not vary with the turnover or velocity. There are figures in the United States on a monthly basis covering approximately fifteen years that show the velocities at which the demand deposits of member banks in some 141 cities have circulated.[1] These figures vary widely from city to city and from time to time in each city. As previously noted, they are high when business is active and when the confidence of business men in the economic situation and prospects is strong. They are low in times of business depression and when the business public is full of doubts as to the economic outlook. In times of boom or of crisis and business depression, they move up and down with the movements of business confidence.[2] Deposits that are not moving are similar to money that is hoarded so far as their influence upon prices is concerned. The following chart shows the movements of velocities for New York City, for 140 important cities exclusive of New York City, and for these 140 cities and New York City combined. The fluctuations during the last seven years have been enormous.

[1] Since the " bank holiday " of March, 1933, the figures cover only 131 cities.

[2] For a study of the relationship between business confidence, deposit velocities, and prices, see the author's article on " Controlled Inflation " in *American Economic Review, Supplement,* March, 1934. See also note on p. 151.

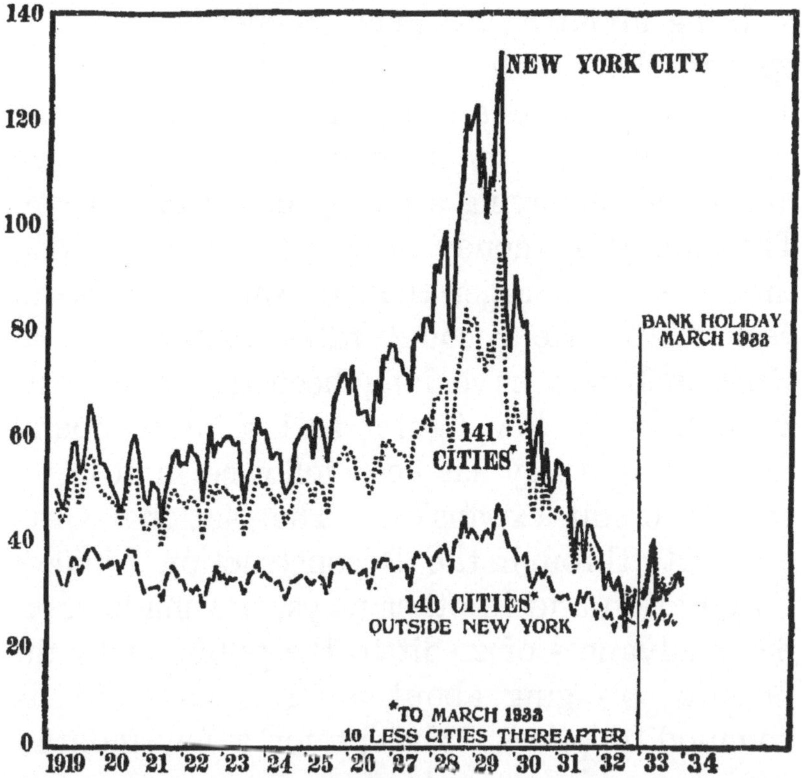

ANNUAL RATE OF TURNOVER OR VELOCITY OF DEMAND DEPOSITS OF MEMBER BANKS IN CERTAIN CITIES

Figures furnished by Federal Reserve Bank of New York City

Efforts to expand the Volume of Money and of Deposit Currency, and thereby to raise Prices, nullified by declining Velocities

Since the crises of 1929 and early 1930, the National Government of the United States and the Federal Reserve authorities have made vigorous efforts to check declining prices and to bring about an advance by means of heavy open-market purchases of Government securities by the Federal Reserve banks, with the intent of glutting the member banks with reserve funds and of forcing a strong credit expansion. The supply of money in circulation for a long time has been larger than it was in the boom year 1929. Re-discount rates at the Federal Reserve banks have long been maintained at an extremely low figure, and a most liberal re-discount policy has been followed in order to enforce credit expansion. The National Government, through the Reconstruction Finance Corporation and in other ways, has made enormous advances of credit to the public, with the hope of bringing about a strong recovery in commodity prices. Except for a few months, however, chiefly in the fore part of the year 1933 and the early months of 1934, business confidence was very low from the collapse of October, 1929, to the summer of 1934, and the velocities of bank deposits moved strongly downward. The efforts at deposit currency expansion were usually more than nullified by the rapidly declining velocities, and, despite all these efforts, general prices continued downward almost continuously from 1929 to the spring of 1933. Since then there has been a modest and halting rise.

Here is an excellent illustration of the difficulty of controlling commodity prices merely by the mechanics of increasing the volume of money in circulation and the volume of reserve bank credit. "You can lead a horse to water, but you can't make him drink." Even if bank credit is cheap and plentiful, responsible business men will not use it if they have little confidence in the business situation and prospect, and if, in their judgment, extending their business gives promise of losses rather than profits. The Government and the Federal Reserve banks may to a considerable extent control the volume of money and of deposit currency in an inconvertible paper money régime, but they cannot control the velocities. These velocities are questions of business confidence and prospects,[1] of hopes and fears, of emotions and prejudices. They are matters of class feeling and of mass psychology. This is the principal reason why in critical times like the present the commodity price level cannot be controlled through regulating the supply of money by changing from time to time the weight of a bullion dollar.

[1] Here we are speaking of confidence in the business situation and prospects. Confidence in a country's money is a different thing. When confidence in business is low, velocities are low, but when confidence in a country's money is very low, velocities tend to be high. Then we have "a flight from money". See pp. 84–86 and 185.

Politics and the Commodity Dollar

Closely related to this obstacle to a successful functioning of a commodity dollar standard is the political obstacle. Commodity prices are affected with tremendous public and class interests. Their ups and downs are matters of vital concern to every class in the community—the farmer, the manufacturer, the business man, the professional man, and the housewife. Taking human nature as it is to-day and American politics as it is, any board at Washington that should try to control the commodity price level in the manner contemplated by the proponents of the commodity dollar would find itself continually under heavy political pressure, both from Congress and the public at large. Under this pressure a scientific and disinterested administration of such a plan would frequently break down. The commodity price index number, which the authorities would be trying to maintain at approximately 100, would be made up of the prices of a larger number of individual commodities. At all times the prices of some of these commodities would be rising and those of others would be falling. Producers of goods whose prices were falling at a time when the general price level was rising would resist by strong political pressure action by the authorities at Washington to increase the weight of the bullion dollar and thereby to push down the price level, which would tend to drag down still farther the already declining prices of their products. What would happen, for example, in a case where the Government's price index numbers were rising while the prices of wheat and cotton, under the influence of unusually heavy crops or of unfavourable markets, were rapidly falling ?

Conclusion

All in all, under present economic and political conditions in America, a price level anchored to a commodity of universal demand, such as gold—a commodity of which there is always an enormous marketable supply, and of which the annual product is but a petty percentage of the world's accumulated stock—is likely to be much more stable and dependable than a price level controlled by any such mechanism as that of the commodity dollar. Such a dollar may work in a model state of a distant Utopian future. It would have a rough passage in the United States of 1934.

CHAPTER X

DEBTOR AND CREDITOR

Present-day American Inflation and its Relation to each—Farm Mortgages of Entire Country are only about Equal to the Real Estate Mortgages of New York City—Who Owes the Billions, and to Whom are they Owed?

THE most important aspect of the present-day inflation problem in the United States is its bearing on the relations between debtors and creditors. The nature of inflation and deflation has previously been discussed and the general principles underlying their relationship to debts have been explained. In this chapter we shall be concerned with the relationship of American inflation, actual and prospective, to debtors and creditors in the United States of to-day.

The Amount of U.S. Debts

What is the volume of present-day internal indebtedness in the United States? An exact answer cannot be given to this question, but there have been a number of estimates made by responsible authorities. These estimates range from $155,000,000,000, the estimate of the National Industrial Conference Board for the interest-bearing debt as of the year 1929, to $285,000,000,000. The differences in the various estimates are due in large part to the extent to which short-term debts, many of them non-interest bearing, life insurance obligations, and bank deposits are included in the estimate. Whatever figure we take, however, it is evident that the volume of American indebtedness to-day is enormous.

Evans Clark in his recently published book, *The Internal Debts of the United States,* made a comprehensive estimate covering all internal debts as of the year 1932, including, in addition to long-term debts, short-term business debts, personal and household debts, and bank debts. His total for all indebtedness was $285,000,000,000, the larger figure cited above. Of this sum, $134,000,000,000 consisted of long-term debts, that is, debts payable over a period of years, but of this total, classified statistics are available for only $126,800,000,000.

It is with long-term debts that we are chiefly concerned in considering questions of inflation and deflation. Most short-term debts are of recent origin, and usually there are no important changes in the value of the dollar between the time the debt is incurred and its maturity. Of course this statement does not apply to periods of strong inflation or deflation.

Included in the $134,000,000,000 of long-term debts was the paid-up value of life insurance policies outstanding, which, in the year 1931, amounted to $17,000,000,000. The face value of these policies, however, amounted to $109,000,000,000.

The greater part of this long-term indebtedness, most of which is in the form of bonds and mortgages, was made by contract specifically payable in United States gold coin of the weight and fineness prevailing prior to the monetary legislation of 1933 and 1934, a contract which the Government has recently made no longer enforceable.

These long-term debts have run for widely varying periods of years. Some were made a generation and more ago and some were made during the present depression. The great bulk of them were incurred between 1913 and 1929. Since the general price level averaged lower during the year 1933 than in any other year since 1916, most of this long-term debt in 1933 and 1934 was payable in dollars of greater purchasing power than were borrowed. In other words, it took more bushels of wheat or of corn, more pounds of beef or of hogs, and more bales of cotton to pay a hundred dollars' worth of debts in 1933 and 1934 than would have been required to do so at the time most of these debts were contracted. This was a serious burden to debtors and continues to be.

Circumstances that Mitigate the Apparent Increase in Weight of the Debt Burden Caused by Recent Deflation

In considering the weight of this increased debt burden and its injustice to the debtor classes, certain mitigating factors should be taken into account that are commonly ignored or underestimated in popular discussion of this subject. They are:

(1) The extent of the injustice varies with the date on which the debts were incurred. All debts, for example, incurred prior to the year 1917 are now and have been throughout the entire depression payable in less valuable dollars than those in which they were incurred. On the other hand, the burden is exceptionally heavy for existing debtors who incurred their long-term debts during the war and early post-war years of inflated prices.

(2) During the stress and strain of the depression, there have been a large number of concessions on the part of creditors to their debtors. Interest rates have been voluntarily

reduced. Reductions in the principal have been granted and debt charges have been permitted to accumulate on a large scale, while many creditors have refrained from making foreclosures in cases of continued default. Some of these concessions have been made under force of necessity and some have been voluntary, but the fact is that they have been made. Of course, a large volume of pre-depression, long-term indebtedness has already been wiped out by bankruptcies. Of a total of 631,439 bankruptcies in the United States during the fourteen years 1920–33, 58 per cent. reported no realizable assets, and the average amount paid to creditors, after deducting costs, was only 7·56 cents per dollar of liabilities.[1]

(3) From 1896 to 1929 (unlike the twenty-three years preceding 1896 when prices were continually falling) the price level in the United States was advancing the major part of the time. In fact, there was only one serious decline, namely, from the summer of 1920 to the summer of 1921, during the entire period, 1896 to 1929. In other words, as previously pointed out,[2] during most of those thirty-six years the majority of debtors were paying their

[1] Starweather and Valenta, *Five Cents on the Dollar*, *Barron's*, May 14, 1934.

[2] See pp. 99 and 101.

debts, both principal and interest, in less valuable dollars than they borrowed. The debtor was benefiting by the instability of the dollar and the creditor was suffering.

(4) A great part of the burden imposed upon the debtor by the high value of the dollar prevailing during the present crisis will in a large proportion of the cases be but temporary if the present depression passes within a short time and the commodity price level returns to about where it was during the eight and one-half years of comparative stability which preceded the crisis of 1929. For reasons discussed elsewhere,[1] I believe that the evidence is strong that the present great dip in the index number of commodity prices is not due to enduring causes any more than was the great rise in commodity prices during the World War and immediately thereafter, and I believe that the 1926 commodity price level in terms of gold would be restored in time, even if there should be no revaluation of the gold dollar. If that opinion should prove to be sound, the principal and interest of the long-term debts that bridged the depression would again become payable in

[1] See Kemmerer, E. W., "The Gold Standard and the Present Economic Situation," in *Proceedings of the American Philosophical Society*, Vol. 71, No. 3 (1932), pp. 87–93 and 97–102.

dollars of approximately the value existing at the times that the majority of the debts were incurred.

Who are the Debtors?

Who are the debtors? Before answering this question, the commonplace fact should be noted that almost everyone is both debtor and creditor, and that it is impossible to divide a population into two distinct classes. In some kinds of business this status of being both debtor and creditor is found on a large scale. A bank, for example, is a creditor to all of its borrowers and to the obligors on all of the bonds it owns, and is a debtor to all of its depositors. A life insurance company is a creditor for all the bonds and mortgages it owns, and is a debtor to all of its policy-holders. If one considers short-term debts as well as long-term debts, there are many more creditors in the United States than debtors.

Roughly speaking, on the basis of the figures given by Evans Clark,[1] the principal debtors in the United States on long-term obligations may be classified, so far as figures are available, in the order of their indebtedness, as follows:

[1] *The Internal Debts of the United States*, p. 6.

AMERICAN DEBTORS ON LONG-TERM ACCOUNT

Debtors	Amount $ Billions	Percentage of a Total of $126·8 Billion
1. Owners of real estate in our cities	27·5	21·7
2. Financial corporations, including life insurance companies (paid-up value of life insurance), investment trusts, and loan agencies . .	21·9	17·3
3. The people of the respective States, cities, towns, etc., owing State and local debts .	18·7	14·7
4. Railways	14·3	11·3
5. All of us, namely, the American public as a whole, who owe in the form of long-term Federal Government debts .	14·2	11·2
6. Public Utilities Corps . .	11·3	8·9
7. Industrial Corps . . .	10·4	8·2
8. Farmers (mortgages) . .	8·5	6·7
	126·8	100·0

It is a significant fact that the smallest item in the classification is farm mortgages, which, in fact, were just about equal for the entire United States to the real estate mortgages of New York City or to less than a third of the urban mortgages for the country as a whole. The largest single class of long-term debts, however, was that of the corporations (railway, public utility, and industrial), which constituted $36,000,000,000, or over a quarter of the total. It is the stockholders of the corporations who owe these debts, for they are the owners of the corporations.

Who are the Creditors?

Who are the creditors to whom these debts are owed? They are, of course, the people, the business concerns, and the institutions who own the bonds, mortgages, and notes. Unfortunately, figures are not available to make possible a comprehensive estimate of the amount owed to the various classes of these creditors. A few significant figures, however, may be cited, based chiefly on the studies of Evans Clark.

The great American life insurance companies, which at the end of the year 1931 had ordinary and industrial policies in force to the number of 122,000,000—equivalent to one policy for every man, woman, and child in the United States—and amounting to $109,000,000,000, have most of their funds invested in bonds and mortgages. They are the largest holders of farm mortgages, with 23 per cent. of the total outstanding in 1928 in their possession; and they held in 1931 over 19 per cent. of all urban mortgage debts. The 641,000 veterans having Government insurance in force to the amount of $3,000,000,000 in 1932 must look to United States Government debt, which is equivalent to Government bonds, as the asset to support this insurance, all of which is payable in a fixed number of dollars regardless of the value or purchasing power of the dollar. Farmers themselves owned about 14 per cent. of all farm mortgages.

Then come the investments of the great savings institutions, such as the mutual savings banks, with nearly thirteen millions of depositors and approximately ten billion dollars of deposits in 1933; the postal savings banks, with over two and a third millions of depositors and nearly twelve hundred millions of dollars of deposits; the building and loan associations, with their ten million members representing the thrifty American home-building and home-buying class. Mutual, stock, and postal savings banks alone held in 1932 ten billion dollars, or nearly all of their investments, in real estate mortgages and bonds; and the building and loan associations held nearly eight billion dollars in urban mortgages alone in 1929, which represented the great bulk of all their assets. The State and national banks of the country, with their savings and time deposits amounting to fourteen billion dollars in 1932, held a large proportion of the assets supporting these deposits in the form of bonds and mortgages and of notes secured by them.

America has long prided herself on the generosity of her people, who by their benefactions have established and endowed her great universities, hospitals, research laboratories, public libraries, and other welfare foundations. A recent study places the value of the property and endowments of these public welfare and religious institutions, which are being administered by trustees, at over nine milliard dollars.[1] By far the largest part of their endowments are in the form of investments payable in a fixed number of dollars, most of which are bonds and mortgages.

These, then, are the principal creditors in America on long-term account—life insurance companies, building and loan associations, savings banks and other banks that take savings deposits, and the great educational, philanthropic, religious, and other welfare institutions.

The largest volume of debts, on the other hand, owed on long-term account by any single class is made up of the bonds which the large corporations owe. Here the real debtors are the stockholders of the corporations.

Investment of Trust Funds largely restricted to Bonds and Mortgages

Furthermore, these life insurance companies, savings banks, building and loan associations,

[1] Wood, Struthers & Co., *Trusteeship of American Endowments*, . . .

and these great educational, scientific, and public welfare institutions have little choice as to whether they will invest their funds in bonds and mortgages, on the one hand, or in so-called equities, on the other hand. The trustee laws of the various States and the banking laws, national and State, in the worthy effort to protect the beneficiaries of trust funds, bring heavy pressure and, frequently, compulsion upon these institutions to invest their funds chiefly in bonds and mortgages; and this pressure is enforced by a strong public opinion. For some time the Government pressure to induce banks to invest heavily in United States Government securities has been strong, and a large and rapidly increasing percentage of the assets of American banks is being so invested.

Inflation in its Relation to Debtor and Creditor

What happens to these creditors under inflation? Obviously, if the debtor pays his debts in a cheaper dollar, the creditor receives payment in this same cheaper dollar. The values of bonds and mortgages in terms of purchasing power decline as the value of the dollar in which they are payable depreciates, and with this decline in the value of the dollar, the value of all life insurance policies, building and loan association credits, savings deposits, pensions, annuities, and endowments of public welfare institutions correspondingly decline. If the great public welfare endowments are destroyed or even substantially reduced by inflation, who will replenish them? They were built up largely by the gifts of wealthy men, during the pioneer years of the United States as a nation. With the country's ever-increasing governmental control of business, and its continually growing resort to progressive income and inheritance taxes for providing public revenues to meet rising government expenditures, will it have in the future the rich men to re-endow these welfare institutions?

Stockholders and Bondholders of Corporations

In the case of corporations, the stockholder tends to receive not only what would be naturally coming to him as the price level rises, but, in addition, what the bondholder loses. This point may be made clear by a simple illustration.

Assume a corporation that has a capital of ten million dollars, of which two million dollars are in the form of long-term, 4½ per cent. first mortgage bonds, three million dollars in the form of long-term 6 per cent. debentures, and the balance of five million dollars in the form of common stock. Assume that this corporation sells an industrial product which wholesales at a net price to the corporation of $1. Assume that the corporation was prosperous before the depression and that its earnings were ample to pay the service on the debt, a fair dividend on the common stock, and to accumulate a modest surplus.

Transferring this corporation now from the prosperous year 1929 to February, 1933, we will assume that the surplus of the corporation has been wiped out, that dividends on the common stock have long been suspended, and that, although the income of the corporation is now much more than sufficient to pay the service on the first mortgage bonds, it is just barely sufficient to meet the service charges on the debentures. Under these circumstances, let us assume that the first mortgage bonds are still selling at 100, that the debentures have gone down to a market price of 50, and that the common stock has declined to 20.

In a situation like this, let us assume that the country experiences a strong inflation, that as a result of this inflation the dollar depreciates to one-fourth of its February, 1933, value, and that the commodity price level, after a considerable period of time, has finally risen to four times what it was in February, 1933.

With this four-fold rise in the price level, the product which the corporation produces, we will assume, has been increased in price from $1 to $4, the manufacturing and distributing costs have risen in the same proportion, and the gross income of the corporation has increased fourfold, with a corresponding increase in the market value of its real estate and equipment. While the corporation's debt in terms of dollars would be the same, it would all now be payable in dollars that would be worth only 25 per cent. as much as the dollars were worth immediately before the inflation. The first mortgage bonds would still be selling at practically 100. The great increase in the income of the corporation would have put it on " Easy Street " as regards its debentures, and these would now have returned to par or better, but they likewise would be payable in " a 25-cent dollar ". The corporation would be receiving four times as many dollars as it did before, and the stockholders' income and property would be four times as large in terms of these less valuable dollars. The stockholders, however, would receive in addition practically all that the bondholder and the debenture holder lost ; because, as the owners of the corporation, they would be paying its entire $5,000,000 of indebtedness (principal and interest) in virtually 25-cent dollars. The common stock, therefore, would rise in price in terms of the new dollars much more than fourfold.

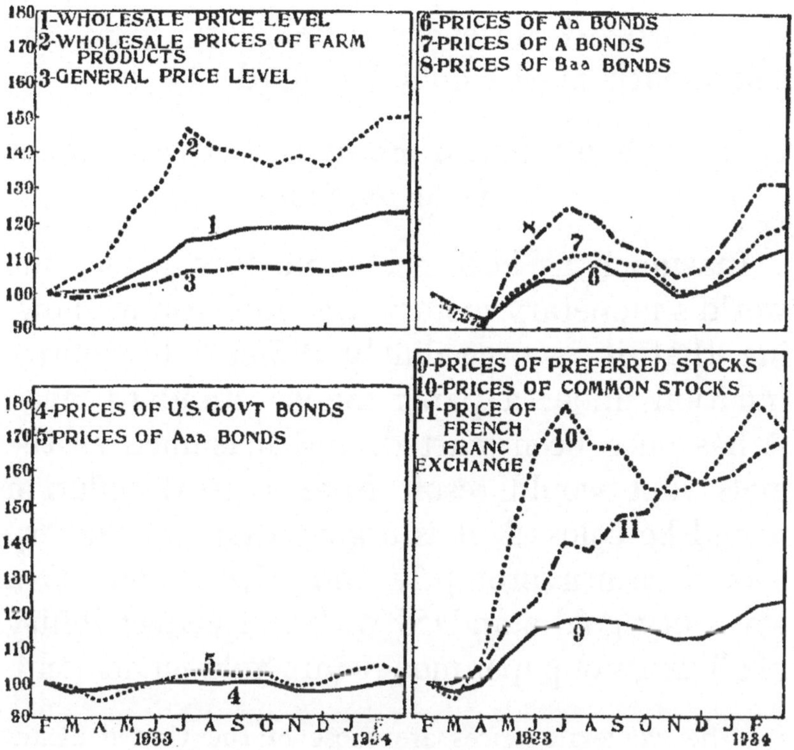

PRICE MOVEMENTS SINCE MONTH PRECEDING DISCONTINUANCE OF GOLD STANDARD IN U.S.
February, 1933 = 100

It is in anticipation of profits of this kind under prospects of inflation that the public shifts its investment demand away from the highest-grade bonds and mortgages, and even from preferred stocks, in favour of common stocks and, to a lesser extent, of the lower-grade bonds.

The chart [1] on page 169 shows in a general way what has already happened in this respect in the United States since February, 1933, the last month of the gold standard.

Country should return promptly and with vigour to Gold Standard

Inasmuch as both economic theory and the world's monetary history are emphatic in showing that it is exceedingly difficult to control inflation under a paper money standard when it has once been started, and inasmuch as the evils that would result from a great inflation would be colossal, it is urgent that the country should return promptly and with vigour to a statutory gold standard with full convertibility of all kinds of paper money into gold on demand.

[1] The wholesale prices are those of the United States Bureau of Labour Statistics; the general price level is based upon the figures computed by the New York Federal Reserve Bank; the figures for Government bonds, preferred stocks, and common stocks are those of the Standard Statistics Company, and those for the corporation bonds are the figures of Moody's Investment Service. All figures are computed on the basis of February, 1933, as 100.

It should be added that the world value of gold has not appreciably changed since February, 1933.

Chapter XI

THE WAY BACK TO GOLD

Monetary Instability is Blocking Prosperity in the United States—Deflation to Old Gold Parity is Politically Impossible—Prompt Stabilization at a Fixed and Permanent Gold Rate would be a Strong Impetus to Enduring Recovery

Monetary Instability an Obstacle to Economic Prosperity

NUMEROUS experiences in the United States and elsewhere have shown that monetary instability, accompanied by widespread fear as to the future value of the monetary unit, is a serious obstacle to economic prosperity. Take, for example, two comparatively recent periods of monetary instability in the United States, the one covering the thirteen years of the depreciated and fluctuating greenback standard immediately following the Civil War, from 1866 to 1878, inclusive, and the other covering the seven years from 1890 to 1896, when the Bryan agitation for free silver was active.

According to Thorp and Mitchell's *Business Annals*, for the first period of thirteen years, four years were prosperous, two were prosperous part of the time and depressed part of the time, and seven were years of depression. For the second period of seven years, one year was prosperous, two were partly prosperous and partly depressed, while four were years of depression. Nearly every country in the world experienced an extensive inflation and great monetary uncertainty during the World War and for a few years immediately after. The world-wide business collapse of late 1920 and early 1921 was one of the prices paid for this inflation. After these experiences the practically universal demand for a speedy return to the gold standard throughout the world, and the actual return of most of the leading countries to that standard during the succeeding eight years, are evidence of widespread dissatisfaction with managed paper money currencies.

The reason why monetary uncertainty is an obstacle to business prosperity is simple. All business looks toward the future. Most business is done on credit and a large proportion of it on long-term credit. Practically all business contracts are in terms of money, and when the currency is being inflated (or deflated), the prices of the various elements that enter into any business, such as the prices of raw materials of different kinds, the prices of machinery and of building materials, the wages of labour, and taxes respond very unevenly and with widely different degrees of lag to the inflation or deflation process. The extent of these responses, moreover, and the time they take place cannot be predicted with even a moderate degree of certainty. Under conditions of inflation, few responsible business men want to make positive commitments for the future, and few people with money are willing to lend it on long-term account, through the purchase of bonds or otherwise, so long as there is uncertainty as to the value of the dollar in which these debts will be paid in the distant future.

At times when the prospects of inflation are strong, as they were in the United States in the late spring and summer of 1933, there may be a feverish business activity in certain fields owing to the anticipation and discounting of rising prices, but such "prosperity" is not enduring, as Americans later learned to their sorrow. Not until the country can establish firmly a monetary standard in which the people upon whose initiative it must depend for economic recovery have strong confidence, can it expect an enduring and orderly prosperity. Putting this idea bluntly, it may be said, the sooner the United States returns to a definite gold standard, with gold convertibility of all forms of paper money on demand, the better.

Any Stabilization Plan should inspire strong Public Confidence

Any stabilization plan to be successful must inspire strong public confidence. There should be no qualifications, no " ifs " or " ands " about it. It should be definite, positive, and permanent. If the Government can stabilize *de facto*, it can even more easily stabilize *de jure*, because the public will not have full confidence in any stabilization plan if it sees that the Government itself is wavering and is not willing to take the responsibility of committing itself positively and permanently to carrying the plan through.

The Way to Stabilize is to Stabilize

The frequently heard talk about waiting until a currency reaches " its natural level " before stabilization is a fallacy. It has been my privilege to assist eleven different countries in stabilizing their currencies on a gold basis, and in every one of them there was talk of a " natural level ", but no one has ever shown, so far as I am aware, just what the natural level of prices is under a depreciated paper money standard or how it is to be determined. In the days of the depreciated greenback standard in the United States when there was much discussion of the time for resuming gold payments, there arose a popular slogan, for which Salmon P. Chase was apparently responsible, that " the way to resume is to resume ". A slogan which might well be adopted to-day is : " The way to stabilize is to stabilize."

Arguments of those who favour a Deflation back to the Old Gold Dollar

There are to-day a number of economists and men of prominence in the financial world—men whose judgment on currency matters is worthy of great respect—who believe that the wise policy for the United States to follow is to deflate the currency back to the old gold dollar. This is the policy that was followed with the greenbacks after the Civil War and which Great Britain followed, mistakenly, according to the judgment of most economists, with the pound sterling after the World War. In the case of the United States it required thirteen years to accomplish it, and in Great Britain's case it took seven years. In both cases the process was a very painful one.

The moral argument in favour of a return to the old gold parity is particularly strong. Nearly all of the long-term debts now outstanding in the United States were incurred when the currency was on the gold standard, and in making these debt contracts both debtor and creditor contemplated payment in terms of the historic gold dollar. To avoid any possibility of doubt on this question, however, it was specifically provided in most bonds and mortgages that payment should be made in gold coin of the United States of the standard of weight and fineness existing at the time the contract was made.

Shortly after the Civil War (1868) the Supreme Court, in the case of Bronson *v.* Rodes, had declared that such contracts were valid and were legally enforceable. During the latter years of the depreciated greenback standard and during the period of the silver controversy ending with the defeat of Mr. Bryan in 1896, there was much uncertainty as to the future value of the American dollar and, therefore, during these periods bonds containing the gold clause could be sold to the public on terms that were much more favourable to the debtor than bonds without this clause.

The National Government, as well as the States and municipalities, gradually adopted the policy of including this gold clause in the loan contract when selling their bonds to the public, and they, like the railways, industrial corporations, and private debtors, profited by the incorporation of this clause. For the Government later deliberately to give up the gold standard and then, after a period of depreciated paper money, to stabilize the dollar at a greatly reduced gold value, and thereafter to pay its own outstanding bonds and to authorize other debtors to pay their bonds and mortgages, which contain this gold clause, at parity in the new and depreciated dollar, it is argued, is a positive breach of faith. The fact that the gold dollar was unstable in value was known when the debts were contracted and, as regards that instability, the parties to the contract presumably took their chances.

Inasmuch as it is unthinkable that the public should pay their debts in gold of the old standard of weight and fineness at a time when all their incomes are in the form of a greatly depreciated paper dollar, the only solution of this moral problem, according to some economists, is to restore the dollar to its previous gold value.

In addition to this moral argument, there are two economic arguments that support this conclusion. The first one may be stated briefly.

Devaluation a Dangerous Precedent

How about the security of loan contracts in the future after a precedent has once been established of deliberately reducing the value of the unit in which debts are payable and of nullifying by governmental action specific agreements that were made with the sole object of avoiding losses to the creditor that might otherwise arise from a depreciation in the gold value of the dollar? If such devaluation is effected once, it may be effected again and then again, and each time with less resistance. The precedent is obviously a dangerous one and as a result of it, the country's credit operations, through which it carries on most of its business, are likely to be rendered more difficult and, to the debtor, more expensive, for a long time to come.

Present High Value of Gold may be only Temporary

The second economic argument in support of the policy is the fact previously mentioned, that the present low price level in terms of gold in the United States, or, in other words, the present high value of gold is probably but temporary and is due to a world scramble for gold for virtual hoarding in a period of world crisis and depression. In this connection, it is a significant fact that, had there been no war and had wholesale commodity prices in the United States risen on an average 2·4 per cent. a year progressively from 1913 to 1929, which is the percentage rate of annual rise from 1896 to 1913, the country would have had approximately the same average price level from 1921 to 1929 that it actually did have.

The great fundamental forces which determine the long swings in the world price level have probably not been greatly changed during recent years. The world's population and the distribution of this population are practically unchanged. Human tastes, human wants, and human capacity to labour are essentially what they were five years ago. The world's great scientific and mechanical accomplishments are still with us and its inventive genius is unimpaired. There is no reason to believe that the world's annual rate of increase in the production of basic commodities, a rate of about 3 per cent. a year which Carl Snyder finds has persisted for two generations, has suddenly been changed permanently. Gold production has increased substantially since 1921, and since 1913 the world's supply of monetary gold has increased much faster than the world's production of basic commodities; while the efficiency of gold as the basis of our credit structure is being increased continually through improvements in our currency and banking machinery. If, since our 8½ years of comparative stability in commodity prices ending in 1929, nothing has happened to change fundamentally and permanently the world's supply of gold and circulating credit, and if, likewise,

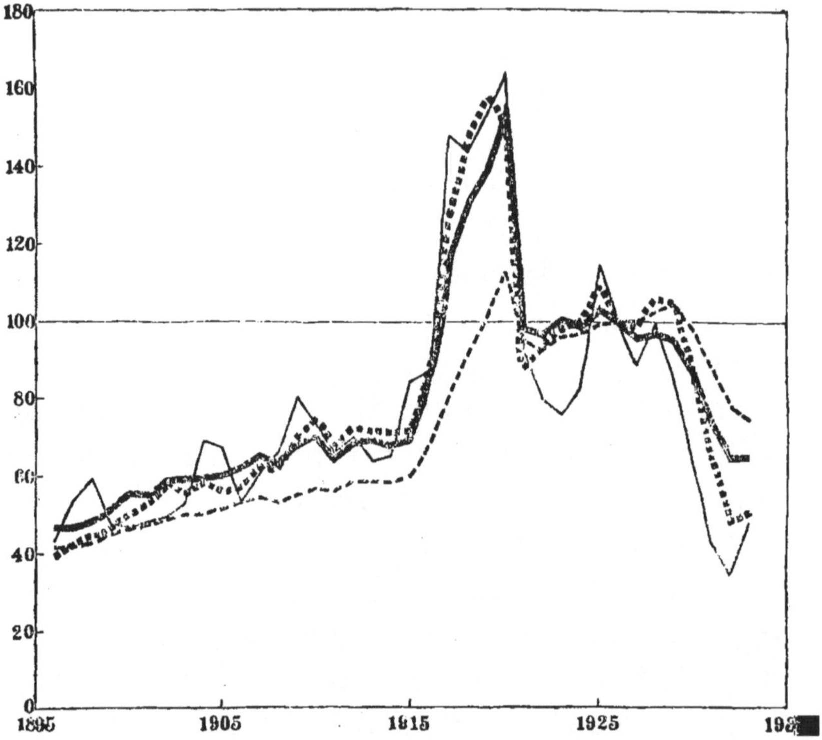

WHOLESALE AND GENERAL PRICE LEVELS AS COMPARED WITH WHOLESALE PRICES OF FARM PRODUCTS 1896–1933

INDEX NUMBERS 1926 = 100

Wholesale price level, B. L. S. Index
General price level, Snyder's Index
Prices of farm products, B. L. S. Index
Prices of wheat in Chicago

nothing has happened to change fundamentally and permanently the world's production of basic commodities and its system of fabrication and marketing, it would seem probable that a commodity price level something like that preceding the crisis would return when we once drag ourselves out of this slough of despond.

The facts just mentioned with reference to the trend of commodity prices since 1896 and to the sudden departure from this trend that took place at the time of the World War and again at the time of the present crisis are shown on the chart, page 180.

Dollar Depreciation Greater in Terms of Gold than in Terms of Commodities

Although the gold value of the paper dollar has depreciated 41 per cent. since the United States gave up the gold standard, which is equivalent to saying that from the standpoint of the value of gold in gold standard countries the United States is now calling 59 cents gold a dollar, and although this depreciation should give the country in time a price level approximately 69 per cent. higher than it had in February, 1933, American prices in general up to the present time have advanced only a minor portion of this percentage.[1] From this it is argued that the depreciation of the dollar up to the present time has been largely in terms of gold and has been due chiefly to the

[1] See pp. 54 and 55 and chart on p. 169.

flight of American capital abroad. On that assumption, it is said that a decision to return promptly and definitively to the old gold standard would cause this capital to return to the United States in large volume, and that this return in itself would go a long way toward restoring the gold exchange value of the dollar to the old gold parity.

Deflation to Old Gold Parity Politically Impossible

These arguments for returning to the old gold parity have great weight. None the less, I believe it would be unwise for the United States to try to return to the old gold parity at the present time, and my reasons for this opinion are chiefly political rather than economic. The opposition to such a policy of deflation would be so strong throughout the country that any attempt to carry it through would delay an effective and enduring return to the gold standard for many years, and during this delay the present depression would be likely to continue to drag on.

The sentiment in the country, both in Congress and out, in favour of inflation is, unfortunately, strong. Widespread agitation in favour of cheap money to lighten the debt burdens of farmers and others has been dangerously effective. People who advocate a reflation of the currency back to the old gold parity are to-day looked upon by a large percentage of the population as Wall Street Shylocks demanding their " pound of flesh ", and even the Wall Street saints are to-day not in particularly good repute in many parts of the country. To safeguard the country against the serious danger now confronting it of a runaway inflation, it is important that the forces of sound money rally on a plan of stabilization that can be made effective quickly. Those who insist uncompromisingly upon returning to the old gold parity at a critical time like the present are butting their heads against a stone wall, heads that could be used in a much more effective way in the fight for sound money.

The political doctors in Washington have for some time been drugging the economic organism with the habit-forming drug, inflation, and are continuing to do so. The patient has resisted and up to the present time the drug has been only mildly effective ; but the patient's system is full of the drug and it cannot be pumped out. Many people have lost confidence in the currency, as shown by the flight of capital from the country, the rise in the prices of common stocks, the slackening demand for high-grade bonds, and the instability in the markets. The giving up of the gold standard in the spring of 1933 was unnecessary and was a great mistake, but Humpty Dumpty has fallen off the wall and it would be an exceedingly difficult and long-drawn-out task to put him together again.

The United States is to-day confronted with the serious danger of a further flight from the dollar, which might end in a runaway inflation. The only certain way to avoid such a runaway inflation and to assure an early return to the gold standard is to stabilize promptly and permanently.[1] A rate of about $66\frac{2}{3}$ cents, which is not far from the present gold value of the paper dollar and which represents about the same proportion of gold parity as that maintained by the paper pound in England since September, 1932, would probably be the best one under the circumstances. In fact, it is very doubtful if a higher rate would now be politically possible.

[1] Since early January, 1934, when this chapter was originally published, there has been a statutory stabilization at a gold value for the dollar of between 50 and 60 cents of the old gold dollar, and an administrative stabilization for the time being at approximately 59 cents. This is described in Chapter XII. While the stabilization law of January, 1934, has apparently eliminated the probability of a future stabilization at any figure above 59 cents, the principles here discussed in connection with the recommendation then made of a $66\frac{2}{3}$-cent rate have in nowise been affected by that legislation.

The value of gold as measured by its purchasing power in gold standard countries has not appreciably changed since the United States gave up the gold standard. The 50 per cent. rise in commodity prices over the level of February, 1933, which the $66\frac{2}{3}$-cent rate might reasonably be expected ultimately to bring about, would give the country a wholesale price level 10 per cent. lower than that of 1926, a general price level 9 per cent. higher and a cost of living 4 per cent. higher. This rate in due time would probably result in an advance of prices more than sufficient to give the debtor classes the relief which the Administration has been advocating. Personally, I believe that within a few years the advance in prices would actually be much further than these figures would suggest, because, for reasons discussed elsewhere,[1] I expect to see gold itself depreciate at least to its 1926 value before 1940. In that case the price level with a $66\frac{2}{3}$-cent dollar would be more than twice as high as it was early in 1933.

Inasmuch as commodity prices will nearly always lag well behind the price of gold on the rise when the paper dollar is depreciating, if the inflation process should be continued until commodity prices themselves reached the 1926 level, and the final stabilization should be effected at

[1] Pp. 159 and 178–181.

the gold value of the paper dollar prevailing when that commodity price level was reached, the final commodity price level would probably be very much higher. The reason would be that the lagging commodity prices would continue to rise for some time after the stabilization and until " they caught up to gold ".

About Two Billion Dollars of Stabilization Profits would go to the Government

If this plan of stabilizing at a gold value of $66\frac{2}{3}$ cents were adopted, every dollar of gold owned by the United States Government and every dollar owned by the Federal Reserve banks would become a dollar and a half, and the approximately four milliard dollars now so owned would accordingly become about six milliard dollars of the new money. Under any plan that may be adopted for reducing the gold content of the dollar in the United States, it is clear that all the profits that are realized should go to the Government as the representative of the entire people.[1]

On the other hand, *if the Government should use these profits as a basis of currency and credit expansion for meeting greatly increased public expenditures—and the political pressure to do so would be heavy—the result would be further great*

[1] See pp. 193 and 194.

inflation and another breakdown of the gold standard. The profits should go to the Government, but the gold should go to the Federal Reserve banks. These profits the Government should use for reducing its indebtedness to the Federal Reserve banks, which at the end of 1933 amounted to nearly $2,500,000,000, as represented by the Government securities owned by these banks. This indebtedness of the Government to the Federal Reserve banks is at present altogether too large, and makes the circulating credit of the country depend unduly upon Government credit, which is greatly influenced by politics. It would be a real service to the American currency and banking system if this debt could be paid off, and if the Federal Reserve banks could become again what they were originally intended to be —central banks devoted to the maintenance of a sound currency and to the provision of reserve credit, when and as needed for the orderly conduct of American business. They were not intended to be, as they are fast becoming, primarily fiscal agencies of the National Government.

Gold Bullion Standard Recommended

Under the plan proposed for a prompt return to the gold standard, the gold bullion standard should be adopted in place of the former gold coin standard. Gold coins are a monetary luxury under existing conditions, and there is no reason why they should be kept in circulation; at least not unless or until a time is reached when gold is depreciating, and it becomes desirable in the interests of monetary stability to put gold coin into circulation again in order to create an increased demand for the metal. Redemption should be on demand in gold bars, and the Federal Reserve banks should not only give gold for notes on demand, but equally should give notes for gold. This latter provision, incidentally, would prevent a sudden rise in the foreign-exchange value of the dollar, that might otherwise result from a heavy return flow of American capital from abroad, upon the official announcement of a strong stabilization plan.

If the Government should come out clearly and positively for such a plan of stabilization, and if the President in a vigorous, Grover Cleveland type of message should declare it to be the intention of the Administration to use all the resources of the Government for maintaining the parity of the currency at this new rate, there would be no danger of a serious depletion of the gold reserves. These reserves would be large, with the increase resulting from the revaluation of the dollar. Europe could not drain them away because she has long since withdrawn most of her liquid funds from the American market. In fact, a heavy return flow to the United States of American capital that has fled abroad to escape the dangers of inflation might well cause a strong temporary flow of gold from Europe to the United States. The gold would not be demanded by Americans for hoarding in their own country or for investment abroad because the stabilization policy of the Government would inspire the public with confidence in its currency. The result of such a bold and prompt stabilization would be a revival of business confidence and a strong impetus toward an orderly and enduring economic recovery.

The President's Stabilization Plan

Just as the first edition of this book was coming off the press, January 15, 1934, President Roosevelt in a message to Congress announced a stabilization plan, and that plan was incorporated in the Gold Reserve Act of 1934, which will be the subject of the next chapter.

CHAPTER XII

THE NEW AMERICAN DOLLAR

Government receives Two Billion Eight Hundred and Six Million Dollars of "Devaluation Profit"—Takes Possession of all Monetary Gold—Law does not establish a Gold Standard—Will probably ultimately Result in more than Doubling Cost of Living—Weakens Federal Reserve System by Transferring many of its Powers to the Government

THE "Gold Reserve Act of 1934" and the President's Proclamation accompanying it, which became effective at the end of January and embodied the recommendations contained in the President's message of January 15, gave the United States a new monetary unit and a new currency system, both of which are of types substantially different from any ever before known throughout the world's widely varied monetary history. Here we have a striking example of bold experimentation.

In this monetary plan there are five outstanding provisions, each of which may be summarized briefly. They are:

1—Statutory Stabilization at a Wide Range of Gold Values

A statutory commitment to stabilize the currency on a gold basis within a range of the gold values represented by 50 to 60 cents of the former gold dollar, supplemented by an administrative order to stabilize, for the time being, at the equivalent of 59·06 cents—a rate which raises the dollar price of an ounce of pure gold from the former statutory price of $20.67 to a new administrative price of $35, representing an increase of 69·3 per cent.

Prior to the discontinuance of the *de facto* gold standard in March, 1933, the value of an ounce of pure gold and the value of $20.67 of all kinds of American money were the same. In practice, even if not required by the strict letter of the law, all of the dozen different kinds of American money were interconvertible on demand to the public with gold at this rate.[1] The American market for gold, moreover, was a free one. Anyone had the right to possess gold without limit and to export and import it without restriction. This was the gold standard. Under the new monetary law, no United States gold coins will be minted

[1] See pp. 5–9.

and none will be permitted to circulate, or even to be held by the public. Convertibility of paper money into gold is entirely at the option of the Government, and, like the exportation and importation of gold, will be permissible only under the conditions and to the extent that the Secretary of the Treasury may determine by administrative order.

2—Government Realizes a Large Devaluation Profit

When the stabilization law was passed, the National Government and the Federal Reserve banks together owned a little over four milliard dollars of the old gold coin and of gold bullion. Since under the devaluation plan the amount of pure gold contained in 59·06 cents' worth of old gold coin is to represent, for the time being at least, the dollar of the new money, this gold will be increased in terms of the new dollars by about $2,806,000,000. Inasmuch as this "profit" arises as an incident of the nation's monetary reorganization, and is in no way the result of services rendered by the Federal Reserve banks or the member banks to the public, the provision of the law that this "profit" should all go to the Government, which represents the whole nation, is entirely sound. In fact, under the Federal Reserve Law as it has stood from the beginning, the stockholding member banks have never had a right to any profits in excess of 6 per cent. cumulative dividends on their stock, and the balance, either currently or at the time of the bank's liquidation, according to law, would always have belonged to the Government. The policy of transferring to the Government all the profits realized upon a monetary devaluation has been the usual one in modern currency reforms, and in recent times has been followed by many countries, including France, Poland, Belgium, Ecuador and Peru.

3—Government takes over Ownership of all Monetary Gold

A third noteworthy provision of the new law is the one declaring that legal title to all gold owned by the Federal Reserve authorities shall be transferred to the United States Government and shall be paid for by the Government in so-called "gold certificates". This policy is supported by a dubious legalistic argument which has been long dear to the proponents of cheap money in the United States, but little recognized elsewhere as regards paper money issues, the argument that the issue of money is a function of sovereignty, and, therefore, should be made directly and exclusively by the Government, a corollary being that the gold reserve should be owned by the Government.

The " gold certificates " with which the Government is paying the Federal Reserve banks for their gold, although carrying the same name as the time-honoured " yellow backs ", which since 1863 had been virtual warehouse receipts for gold coin and until the spring of 1933 were payable to bearer in gold coin on demand,[1] are of a very different character, and to call them " gold certificates " is misleading. It is true that they are " backed " dollar for dollar by gold bullion. They are not, however, convertible into gold bullion to bearer on demand and are not expected to circulate. They are to be held only by the Federal Reserve authorities and the Government. They are redeemable in gold only at the option of the Government and to the extent considered necessary by the Government for " settling international balances " and for maintaining the " equal purchasing power of every kind of currency in the United States ". No possessors of money, not even the Federal Reserve banks with their milliards of dollars of gold certificates, now enjoy a statutory right to have their money converted into gold on demand. Con-

[1] See pp. 18 and 19.

vertibility in the future will be entirely at the option of the President.

4—Commodity Dollar Authorized

A statutory provision permitting the President to vary the price of gold within such limits as to establish a commodity dollar that would have a gold value varying between 50 and 60 cents of the old gold dollar is a fourth important item in the new monetary system. The fixing of an upper limit to the range for possible stabilization was apparently necessary to enable the Government to establish its devaluation profit, for how, otherwise, with merely a lower limit for devaluation set by law, could it know what would be the minimum profit which it could take over, or even that there would ultimately be any profit at all? The granting to the President of authority administratively to fix the gold value of the dollar from time to time at any point within such a wide range may have been motivated in part by the desire to strengthen his position for bargaining with other countries in connection with future plans for international monetary stabilization. The main motive, however, appears to have been to authorize a commodity dollar standard. As it is, the President now has, in addition to his practically unlimited legal authority to experiment with bimetallism, authority to experiment at will with the commodity dollar within a range of a gold value of $16\frac{2}{3}$ per cent. He may in his discretion vary the official price of gold anywhere from \$34·45 to \$41·34 an ounce.

5—Wide Discretion granted to President in use of Devaluation Profits

A fifth important feature of the new monetary law is the granting of wide discretion to the President as to the uses to which he may apply the \$2,806,000,000 of " profits " realized from the reduction in the gold content of the dollar. Two milliard dollars of this sum are to constitute for a period of two or three years a so-called " stabilization fund " " available for expenditure under the direction of the Secretary of the Treasury, and in his discretion, for any purpose in connection with carrying out the provisions of this section, including the investment and reinvestment in direct obligations of the United States of any portion of the fund which the Secretary of the Treasury, with the approval of the President, may from time to time determine are not currently required for stabilizing the exchange value of the dollar ".

This apparently carries authority to use the fund to prepare and maintain a market for the heavy bond issues which the Government is now contemplating.

None of the Recently Enacted Hodge-podge of Emergency Monetary Laws are Repealed

Unfortunately, the new law continues all the extravagant hodge-podge of monetary provisions contained in the so-called "Thomas Amendment" of May 12, 1933, and the Joint Resolution repealing the "gold clause" passed June 5, 1933.

Let us consider briefly a few of the principal characteristics of the currency system which the new law will provide.

What is the New Monetary Standard?

The standard created by this new legislation is difficult to define, for such a standard has never before existed. Legally, it is probably best classed as a restricted commodity standard, for the monetary unit fixed by law may have a range in gold value anywhere from 50 to 60 per cent. of the value of the former gold dollar; and the law apparently contemplates the possibility of varying the gold value of the dollar within this range according to the ups and downs of the commodity price level. In practice, the new currency might be so administered as to become a *de facto* gold standard.

So long as the Government buys and sells gold on demand at approximately $35 an ounce, and maintains a free market for gold both within the country and in foreign trade, the value of the paper dollar will be maintained at practical parity with the value of a fixed quantity of gold in a free market, and this is the "constituting quality" of the gold standard.

To the extent, however, that the Government interferes with the free market for gold or exercises its legal authority to vary the gold content of the dollar by altering the buying or selling price of gold beyond the limits of the so-called "gold points", the value of the dollar will tend to depart from the value of a fixed quantity of gold in a free market and the gold standard will cease to exist. The country would then, actually as well as legally, be on a restricted commodity standard if the bullion content of the dollar were raised and lowered to compensate for advances or declines in the commodity price level, and it would be on a restricted fiduciary paper money standard if there were no definite tie-up of the dollar with the commodity price level.

The Future Price Level

The principal motive advanced for devaluation has been the raising of commodity prices for the purpose of lightening the burden of persons carrying long-term debts. The advocates of devaluation have usually favoured a reduction in the gold content of the dollar sufficient to restore the commodity price level of 1926, which was, broadly speaking, about the level prevailing from the summer of 1921 to the end of 1929.

Speaking in terms of the former gold dollar, it is obvious that if the Government should call the amount of gold represented by one-tenth of a five-dollar gold piece a dollar, commodity prices in time would be twice as high as they would be if it should call the amount of gold in one-fifth of a five-dollar gold piece a dollar. For the same reason, if the United States stabilizes at 59·06 cents of its old gold dollar, calling the value represented by this amount of gold the new dollar, sooner or later commodity prices will rise to a level about 69·3 per cent. higher than they would have been had the old gold dollar been retained. This rise in prices, of course, would take time, and the advance would not assume much importance until in some way or other an increased demand for goods and services should be created; for the mere reduction of the gold content of the dollar and the mere pumping of more money and bank credit into circulation will not of themselves raise prices.[1] Either procedure might, in fact, reduce confidence and thereby reduce the velocities of circulation of money and bank deposits so much as to cause a temporary decline in prices. However, sooner or later a reduction in the gold content of the dollar, through its influence on foreign trade, on the international flow of gold, on the gold-mining industry, on the relations between debtors and creditors, on interest rates, and in other ways, will set into motion powerful economic forces that will make commodity prices higher than they otherwise would have been, and ultimately the price level will register the full reduction in the gold content of the dollar. Whenever a devaluation is made, however, repeated experience has shown that advances in commodity prices are very uneven. Wholesale prices in general rise faster than retail prices, and retail prices faster than wages. Within each class of prices the rates of advance are very unequal for different kinds of commodities and services. Even with pronounced changes in the metallic content of a monetary unit, it usually takes many years before a complete adjustment to the new equilibrium has been made.

Bearing these facts in mind, we may say

[1] See pp. 145–51.

that the new American gold dollar of 15·24 grains of gold, ·900 fine, would ultimately give the country a general price level approximately 69·3 per cent. higher than it would have had under the old dollar. But what would have been the price level under the old gold dollar if the United States should have ultimately worked its way out of this depression without devaluation? No one can answer this question with certainty. The factors involved are too many and too complicated. On the basis of certain assumptions, however, we may form a general idea of what might have happened.

Future Commodity Prices, assuming Continuance of Present High World Value of Gold

If, for example, we should accept the contention of those economists who hold that the commodity price level of the years 1921 to 1929 in gold standard countries was being artificially held up to an abnormal height by over-extended bank credit, a delayed return of many countries to the gold standard after the war and the widespread use of an "unworkable and over-expanded gold exchange standard" in lieu of a gold metal standard, and if we should agree with these economists that the world is suffering from an enduring scarcity of gold, we might accept the low commodity price level of February, 1933, namely, of the month before the United States gave up the gold standard, as a normal price level under present-day conditions for a currency based on the old American gold dollar. In other words, we might assume that the low commodity price level of early 1933 had come to stay for a long time if the old gold standard and the old gold dollar had been retained. In February, 1933, wholesale prices in general were 40 per cent. lower than for the year 1926, general prices were 27 per cent. lower and the cost of living was 31 per cent. lower. Since the value or purchasing power of gold in gold standard countries remained practically unchanged from February, 1933, to the spring of 1934, if we increase the commodity price index numbers of February, 1933, by 69·3 per cent., the exact percentage necessary to compensate for the Government's reduction in the gold content of the dollar, wholesale prices would be 37 per cent. higher than they were in the spring of 1934, and 1¼ per cent. higher than they were for the year 1926; the general price level would be 55 per cent. higher than for March, 1934, and 23 per cent. higher than for the year 1926, and the cost of living would be 54 per cent. higher than for April, 1934, and 17 per cent. higher than for the year 1926. Even on the assumption, therefore, that the country has been suffering from a great and enduring scarcity of gold, " the new 59·06 cents gold dollar " would probably give the United States in time a general price level and a cost of living far above those of the present time and well above those of 1926.

Future Commodity Prices, assuming that Gold depreciates to its 1926 World Value

If, on the other hand, we should make what I believe to be a much more reasonable assumption, namely, that the commodity price level of the United States of the year 1926, which represented roughly the level prevailing for eight and a half years ending with December, 1929, and which represented about the level that would have existed had there been no war and had prices moved upward from 1913 to 1929 at the same rate at which they advanced from 1896 to 1913; if we should make the assumption that this level represents roughly what will be the purchasing power of gold when the world once works its way out of this depression and when our panic-stricken international scramble for gold has subsided;[1] then the American commodity price level, when once the adjustment to " the new 59·06 cents dollar "

[1] See pp. 178–81.

has been completely effected, would be as follows: All prices would be about 69 per cent. higher than in 1926. Wholesale prices would be 129 per cent. higher than for March, 1934; general prices would be 114 per cent. higher, and the cost of living would be 124 per cent. higher. *These are the price advances that we may reasonably expect will ultimately have taken place when the depression is over and when the results of the devaluation have completely worked themselves out.*

The above estimates, however, are on the optimistic assumption that the powerful inflationary forces now at work, both political and economic, can be effectively controlled and that the newly established monetary standard can be maintained. If this assumption should prove false, and if inflation should once break from control and we should have a strong flight from the dollar, prices would rise to very much higher figures.

Our New Monetary Legislation and the Federal Reserve System

An important phase of the new monetary law is its probable effect upon our Federal Reserve System. The subject is a large one and only a few high spots can be touched upon here.

The principal functions of a central bank or of

a group of federated central banks, like our Federal Reserve banks, are briefly as follows: (1) To provide the country with a sound and elastic bank-note currency, a currency that will expand as trade demands increase and contract as they decrease. (2) To provide the commercial banks of the country with funds if and as needed to meet emergency demands for currency and credit and the regularly recurring seasonal demands. (3) To hold the country's gold reserves. (4) To regulate and maintain the orderly functioning of the nation's money market, at one time preventing dangerous credit expansion, and at another time preventing undue contraction, and at all times exerting a regulating influence on the exportation and importation of gold. For thus conserving and regulating the nation's money and its money market, a central bank usually has as its principal instruments the exclusive right of note issue, the right to raise or lower official discount rates, and the right to buy and sell certain types of commercial paper and securities in the open market. (5) To act as the depository and fiscal agent of the Government.

The " Gold Reserve Act of 1934 " and other recent monetary and banking legislation and Governmental regulations have so weakened the powers of the Federal Reserve System in the fields in which a central bank usually operates, except the purely Government field, as to prevent it from functioning in a normal and effective way. Its powers have been extensively taken over by the Government and it is no longer fair to hold it responsible for performing the usual functions of a central bank.

The Federal Reserve System can no longer provide the country with an elastic currency, for its notes are increasingly issued to meet the fiscal needs of the Government against the security of the Government debt, while the large powers now enjoyed by the Government over the issue and deposit of gold certificates, the issue of silver certificates, Federal Reserve bank notes and greenbacks threaten to break down completely the control of the Federal Reserve banks over the currency. Furthermore, the member banks are more and more looking to the Government or to Government agencies for funds, and the loans and discounts of the Federal Reserve banks for member banks have for some time been exceedingly small. For example, out of total resources of about eight milliard dollars on May 9, 1934, for all twelve Federal Reserve banks the total amount of bills discounted was only 37 million dollars, and of this amount about 6 million dollars were secured by Government obligations. The total amount of United States Government securities owned on that date by the twelve Federal Reserve banks amounted to 2·4 milliard dollars and represented more than 98 per cent. of the total bills and securities owned. The Federal Reserve banks no longer hold any of the country's gold reserves, and they cannot control the export or import of gold, or even require the redemption in gold of a single dollar of the milliards of dollars of gold certificates they hold. The ownership and control of the gold has passed completely to the Government. The power of the Federal Reserve banks to control and conserve the money market through changes in the discount rate and through open market operations, for the present at least, is nearly gone. The market is so flooded with funds that very few member banks wish to borrow from the Federal Reserve banks, and the flooding has been at the behest of the Government in carrying out its reflation programme, and not on the initiative of the Federal Reserve banks. The operations of the Federal Reserve banks are being increasingly dictated by the fiscal needs and the monetary policy of the Government. If currency and credit redundancy should threaten to bring on a dangerously speculative market, the Federal Reserve banks would find it very difficult if not impossible to carry through a contraction policy by raising discount rates and selling Government securities in the market, in case the Treasury at the same time were carrying on an extensive programme of Government debt flotation, as it seems likely to be doing for some time to come. Such a policy of contraction would hurt the market for Government securities. Furthermore, the Federal Reserve System would be powerless to control the market in the face of the operations of the Treasury Department with its new 2 billion dollar stabilization fund. These operations will almost of necessity dominate the situation.

The Federal Reserve banks may cut bait, but the Government will do the fishing. If present tendencies are long continued, the Federal Reserve System will become little more than a fiscal agency of the Government. Like the First and Second United States Banks, its tenure of usefulness as a real central bank will have been but twenty years.

Danger of Inflating on the Basis of the Devaluation Profits

With the exception of the threatening drastic silver measures, probably the greatest danger immediately facing the new stabilization plan is that the stabilization fund of two milliard dollars may be used largely for fiscal rather than for purely stabilization purposes, and that this fund, together with the other eight hundred and six million dollars of devaluation "profits" may be made the reserve basis for an excessive credit and paper money expansion.[1] If this should be done, it would seriously threaten a breakdown of the newly established monetary unit.

[1] See p. 189.

INDEX

 P*